Energy Storage Devices

Edited by M. Taha Demirkan and Adel Attia

Published in London, United Kingdom

IntechOpen

Supporting open minds since 2005

Energy Storage Devices
http://dx.doi.org/10.5772/intechopen.78502
Edited by M. Taha Demirkan and Adel Attia

Contributors
Sun-Jae Kim, Hyeon-Woo Yang, Dallia Mahmoud Morsi Ali, Mahmoud Elsisi, Malek Belouda, Francis Assadian, Kevin Mallon, Brian Walker, Essamudin Ali Ebrahim, Sangwon Kim, Fuming Chen, Karthick Ramalingam

First published in London, United Kingdom, 2019 by IntechOpen
IntechOpen is the global imprint of INTECHOPEN LIMITED, registered in England and Wales, registration number: 11086078, 7th floor, 10 Lower Thames Street, London, EC3R 6AF, United Kingdom
Printed in Croatia

British Library Cataloguing-in-Publication Data
A catalogue record for this book is available from the British Library

Additional hard and PDF copies can be obtained from orders@intechopen.com

Energy Storage Devices
Edited by M. Taha Demirkan and Adel Attia
p. cm.
Print ISBN 978-1-78985-693-4
Online ISBN 978-1-78985-694-1
eBook (PDF) ISBN 978-1-83880-383-4

For EU product safety concerns:
IN TECH d.o.o., Prolaz Marije Krucifikse Kozulić 3, 51000 Rijeka, Croatia, info@intechopen.com or visit our website at intechopen.com.

Meet the editors

M. Taha Demirkan received his BSc degree in physics from Gazi University, Turkey in 2006 and his MSc degree in materials science from Drexel University, USA in 2010. He finished his PhD study at the Department of Applied Science at the University of Arkansas at Little Rock, USA in 2014. He is currently a research fellow at Gebze Technical University, Turkey. His research is focused on electrochemical energy storage with an emphasis on high energy anode materials for Li-ion batteries.

Adel Attia has worked in many areas related to materials electrochemistry such as electrochemical corrosion, lithium-ion batteries, mesoporous materials, electrochemical and photochemical degradation of some organic compounds, and electrochemical characterization of materials for organic photovoltaic cells. Dr Attia received his Ph.D. from Charles University in collaboration with J. Heyrovský Institute of Physical Chemistry, Prague, Czech Republic in 2003 and joined the State Key Lab for Physical Chemistry of Solid Surfaces, Xiamen University, China from 2005 to 2007. Thereafter, he joined Monash University, Melbourne, Australia from 2008 to 2010. Currently, he is working for the National Research Centre, Cairo, Egypt as a research associate and collaborating with other members to develop a new nanocomposite coating for corrosion protection of some metals and alloys and their electrochemical characterization.

Contents

Preface

Energy storage will be a very important part of the near future and its effectiveness will be crucial for most future technologies. Energy can be stored in several different ways and these differ in terms of the type and the conversion method of the energy. Among those methods; chemical, mechanical, and thermal energy storage are some of the most favorable methods for containing energy. Current energy storage devices are still far from meeting the demands of new technological developments. Therefore, much effort has been put to improving the performance of different types of energy storage technologies in the last few decades. This book explores different methods for energy storage and analyzes the main parameters that affect the utilization of energy storage devices, such as energy density, performance, and cost. Modelling on several hybrid energy storage systems is discussed and this is followed by a proposed control mechanism for energy storage devices. Later chapters include several recent topics such as new advancements in lithium-ion batteries, energy storage systems with redox flow battery, and hydrogen energy storage.

Dr. M. Taha Demirkan
Research Fellow,
Gebze Technical University,
Materials Science and Engineering Department,
Kocaeli, Turkey

Adel Attia, Ph.D.
Associate Professor,
Department of Physical Chemistry,
National Research Centre,
Cairo, Egypt

Section 1

Modelling of Hybrid Energy Systems

Chapter 1

New Design Methodologies for Sizing Electrochemical Storage in Renewable Energy Systems (Case Study: Wind Turbine System)

Malek Belouda

Abstract

This chapter presents four original methodologies for sizing electrochemical storage devices in renewable energy systems. The case study is taken to apply these methodologies on an electrochemical storage device (a battery bank) inside a wind turbine system. The storage device acts together with wind cycles and consumption profile, particularly for a remote application. In general, in a context of optimal design for such systems, the optimization process time (long processing time) is hampered by the wide number of system simulations caused by the long duration of the actual wind speed measurements used as input data for the problem. Two sizing methodologies are based on a statistical approach, and the two other methodologies are based on the synthesis of compact wind speed profiles by means of evolutionary algorithms. The results are discussed from the point of view of the relevance of the battery bank sizing and in terms of computation cost, this later issue being crucial in view of an integrated optimal design (IOD) process.

Keywords: renewable energy systems, electrochemical storage, sizing, wind profile synthesis, optimization, evolutionary algorithm

1. Introduction

Renewable energy productions are characterized by the unpredictability and the intermittence of the environmental data such as solar irradiation and wind speed in photovoltaic and wind system productions. Therefore, the main criteria when supplying remote areas from renewable source (wind energy/solar irradiation) are the continuity and reliability of electricity supply. The satisfaction of these two criteria can be reached by inserting storage devices (electrochemical devices, hydraulic devices, etc.), but the high owning cost of such solution denotes a major inconvenient for this alternative [1–5]. Hence, an optimal sizing design of the renewable production system coupled with the storage device appears as a guarantee to assure reliability and cheap electricity to supply consumers in isolated sites. The optimal design is achieved by performing global optimization process using a several simulations [6–12]. Nevertheless, these simulations are performed in large time duration, since the environmental data (wind speed, solar irradiation) are characterized by unpredictability which needs great amounts of such data.

In this context, this chapter scrutinizes the optimal design of an electrochemical storage device (a battery bank) associated to a renewable energy system (a wind turbine (WT)) in order to supply continuously a typical farm in a remote site, considering environmental data potentials and load demand variations are a crucial step in the design of these systems. In the case study, the battery bank is exposed to a "time phasing" (T_{ph}) between the generating WT energy/power (consequences of the wind data) and the consumption profile with a time cycle of 24 h, which is a specific problem when sizing the battery bank: indeed, the difference between power production and power consumption profiles is not sufficient to characterize the battery sizing. The time phasing of this power difference is also of prime importance as it sets the battery energy which is also essential in the battery sizing process.

Four generic battery sizing methodologies are investigated. Two methodologies are based on statistical approaches, and two other methodologies are based on compacting environmental data duration. These methodologies are applied, as a case study, on a renewable energy system consisting of 8 kW standalone wind turbine (**Figure 1**).

Statistical methodologies determine the power and energy constraints associated with the battery bank from temporal Monte-Carlo-based simulations including environmental data and consumption profile variations. Environmental data evolution is considered as stochastic, while the consumption demand is deterministically day to day regenerated (**Figure 2**). Only slow dynamics of the wind potential is taken into account. This means that fast dynamics of wind speed related to

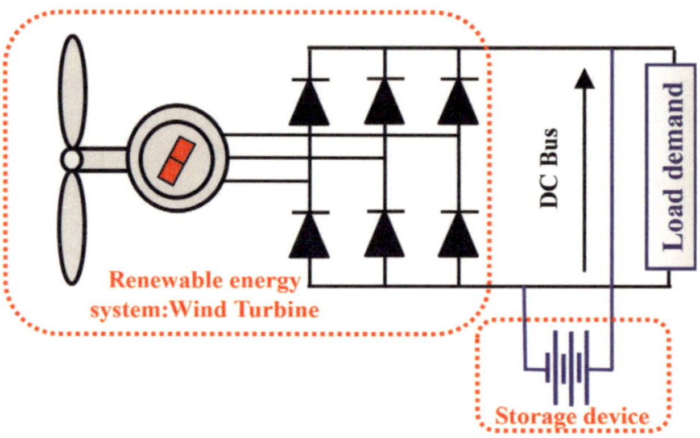

Figure 1.
Case study: a WT system with battery for standalone application.

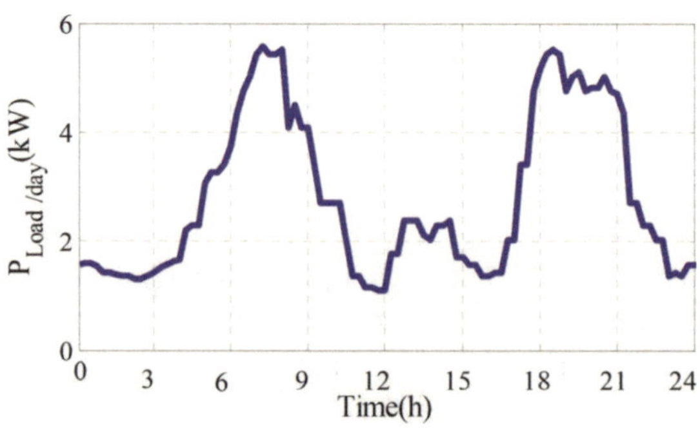

Figure 2.
Daily load demand profile.

turbulence is neglected. Therefore, a Weibull distribution represents wind speed features. To find the most critical constraints on the battery, we require including all correlations between renewable power production and load profile (e.g., time windows with high wind powers and small load powers and inversely). So, the process computation cost is rather expensive especially when a global integrated design process is performed, where all components have to be simultaneously optimized. Thus, the computation cost of these statistical approaches presents an actual problematic. In order to face this problem, two other methodologies are investigated for reducing environmental data profile durations while keeping their feature trace in terms of variability, intensity, and statistics. These approaches are based on the original approach proposed in [13]. This latter approach is adapted for compacting wind speed profiles. The idea consists of aggregating elementary-parameterized segments to generate a compact environmental data profiles. This is performed by satisfying target indicators representing the environmental data features of a reference profile of larger duration. This inverse problem involving the determination of the segment parameters is solved using a genetic algorithm.

2. Renewable system description

The considered system is a 8-kW-full WT battery charger without active control and with minimum number of sensors (**Figure 1**). This WT is sizing in manner that the wind power extraction of this configuration matches very closely the behavior of active WT systems operating at optimal wind powers by using an MPPT control device. The deterministic load profile is set on 24 h and day by day repeated as indicated in **Figure 2**.

A lead acid Yuasa NP 38-12I is considered as a battery element. The basic characteristics are summarized in **Table 1**.

The battery sizing algorithm is based on an upper saturated integration of powers in the battery bank. The idea of a saturated integration of the battery power is related to considering charge powers are no more integrated if the state of charge (SOC) of the storage device reaches its maximal level. Thus, we consider that charge power is wasted in order to avoid the storage device oversizing occasioned with a simple integration especially during huge wind speeds with reduced consumption.

Nominal capacity C_3	30.3 (Ah)
Nominal voltage V_0	12 (V)
Nominal discharge Current I_3	10.1(A)

Table 1.
Basic characteristics of the considered lead acid battery element.

3. Statistical battery bank sizing methodologies

3.1 First statistical approach (environmental data profile distribution)

At a particular location characterized by a specific wind-energy potential, wind speed can be predicted by several statistical distribution models from the wind-energy potential. In this approach, the sizing process is based on the generation of a wind cycle from its statistical distribution [14, 15].

The consumption profile and the power and energy levels which depend on the wind potential magnitude and phase decide the storage device sizing. Hence, determining the pertinent storage device sizing must be under the "worst" case conditions (maximum power and energy). In order to realize these conditions, several wind speed profiles with increasing duration have to be produced until battery sizing stabilization (**Figure 6**), i.e., battery cells become quasi-constant.

The synoptic of the random process of wind speed generation of the is shown in **Figure 3**. The continuous temporal wind speed profiles are generated from statistical distribution by interpolating some number of samples generated with a random number generator according to the recognized statistical distribution.

Figure 4 shows the synoptic battery bank sizing process. The idea consists of generating 11 wind cycles with a progressive duration from 1 to 200 days. These cycles are synthesized from a Weibull distribution of the wind speed during Nd days (Nd = {1, 2, 3, 10, 20, 30, 50, 70, 100, 150, 200 days}). After simulation of the WT system, 11 extracted wind powers (Pwind) are produced. The consumption power (Pload) is daily repeated during the Nd days.

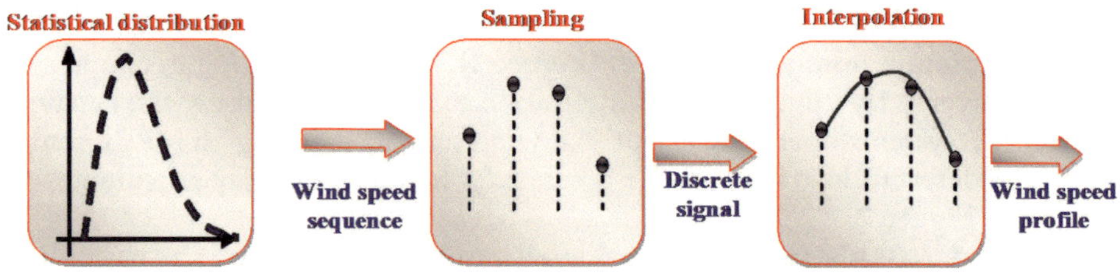

Figure 3.
Wind speed generation process.

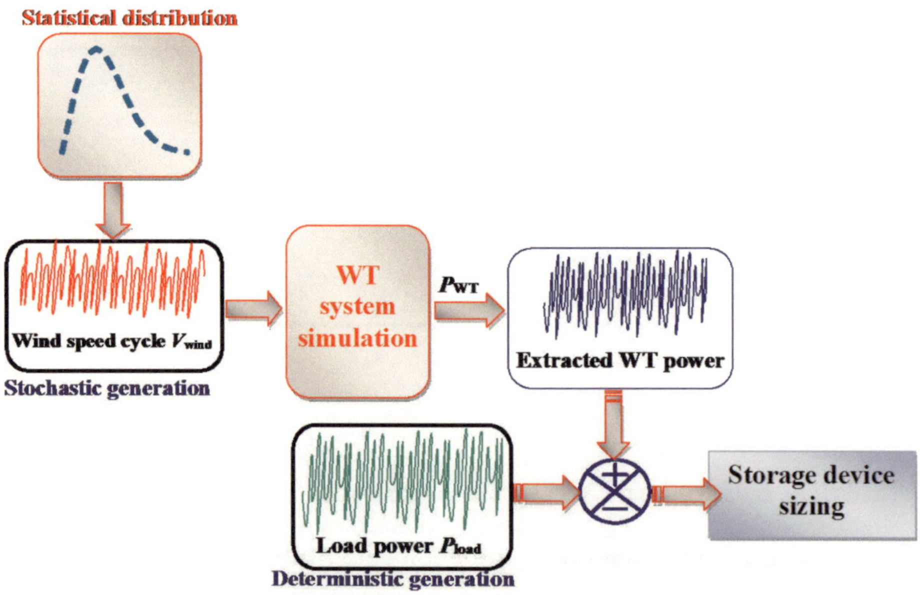

Figure 4.
Battery bank sizing process based on wind profile generation form its distribution.

Note that the battery power used by the sizing algorithm is given by:

Table 2 shows the battery element number and the computational time (T_{CPU}) under different wind speed profiles. The computational time is the time needed by the processor to simulate the system model and to perform the storage device sizing process.

3.2 Second statistical approach (extracted wind power distribution)

In order to reduce T_{CPU}, a critical factor in an integrated optimal design (IOD) context, this approach is based on the direct generation of the extracted power (PWT) histogram instead of the wind speed histogram as proposed in the first methodology. The extracted power for each wind speed interval is estimated by simulating the WT system. PWT is synthesized on the same time scales as with the first methodology.

The PWT histogram is built from wind statistics. Thus, we obtain directly the WT power profile from its distribution by means of random number generation and interpolation techniques exactly as described in the first methodology. Therefore, the PWT can be directly generated before to obtain the battery power PBAT used for the storage bank sizing process (**Figure 5**). Similarly to the first methodology, 11 PWT cycles are produced with a progressive duration from 1 to 200 days and waiting until stabilization of the number of battery cells.

Number of days	CPU Time(s)		$\langle N_{b,r} \rangle$	
	Wind speed-based approach	Output WT power-based approach	Wind speed-based approach	Output WT power-based approach
1	0.1	0.06	46	45
2	0.19	0.17	54	57
3	0.29	0.26	61	61
10	0.56	0.31	88	88
20	0.78	0.32	97	98
30	1.21	0.41	109	114
50	3.06	0.58	117	118
70	4.57	0.83	130	129
100	7.06	1.32	126	127
150	12.8	2.47	126	124
200	19.23	4.00	132	131

Table 2.
Statistical approach results.

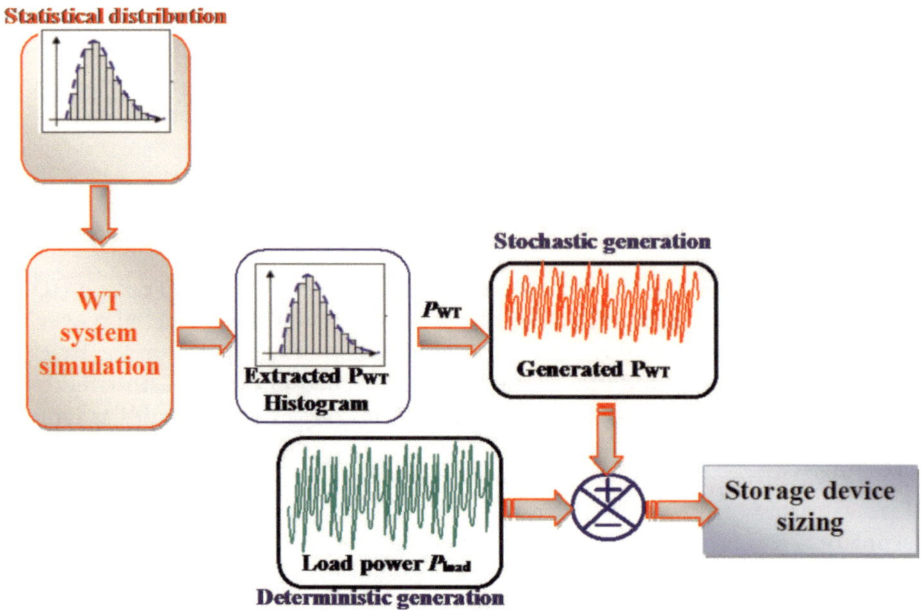

Figure 5.
Storage device sizing process based on extracted power generation profile form its distribution.

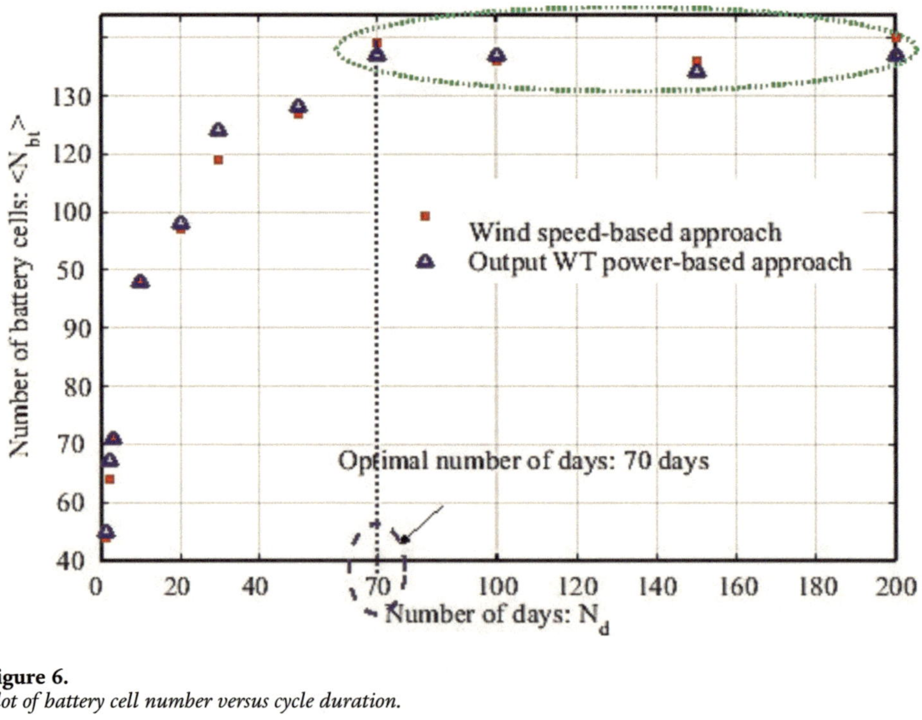

Figure 6.
Plot of battery cell number versus cycle duration.

3.3 Results

To face the stochastic nature of wind speed, several simulations of the 11 wind speed cycles (with an increasing number of days from 1 to 200) have been performed. **Table 2** gives the average number of battery cells $<N_{bt}>$ obtained after 10 simulations for both methodologies. $<N_{bt}>$ obtained from the 11 generated wind speed cycles are shown in **Figure 6**.

4. Battery sizing based on "compact synthesis approach"

In this approach an actual wind speed profile of 200 days duration is considered as reference data. In order to generate a compact wind speed profile with a reduced duration $\Delta t_{compact}$, the "compact synthesis process" is applied on this profile. Two methodologies are scrutinized, differenced by the target indicators used for generating the fictitious compact wind speed profile in order to establish its correspondence with reference actual profile.

4.1 Synthesis process of compact environmental data profiles

The principle of compact environmental data synthesis process consists of the generation of a fictitious profile of temperature, solar irradiation, wind speed, etc. by satisfying some constraints related essentially to variable characteristics, i.e., minimum, maximum, and average values, probability distribution function, etc. These constraints are expressed in terms of "target indicators" that can be evaluated from a set of reference profiles usually of large duration: here we have considered a 200-day wind profile. The fictitious profile is obtained by aggregating elementary segments as shown in **Figure** 7. Each segment is characterized by its amplitude ΔS_n ($\Delta S_{minref} \leq \Delta S_n \leq \Delta S_{maxref}$) and its duration Δt_n ($0 \leq \Delta t_n \leq \Delta t_{compact}$).

In order to fulfill the constraint related to the time duration, i.e., $\sum \Delta t_n = \Delta t_{compact}$, a time scaling step is executed after the variable profile generation. The compact fictitious profile generation synthesis consists of finding all segment parameters fulfilling all target indicators given by the reference data (actual profile) on the reduced duration $\Delta t_{compact}$. This is performed by solving an inverse problem, using evolutionary algorithms, with 2 N parameters where N denotes the compact profile segment number [16]. As evolutionary algorithm we have chosen the clearing method [17] well suited to treat this kind of problem with high dimensionality and high multimodality. Target indicators are also related to the design context itself (in this case study, the WT system has to charge a battery bank for which maximum powers and energy range are pertinent).

4.2 Compact synthesis approach based on storage system features

The first approach uses, as target indicators, the storage system features. The storage system global sizing is related to the maximum storage power P_{BATMAX}, the minimum storage power P_{BATMIN}, and the maximum energy quantity imposed to this storage ES. These variables target indicators of the inverse problem, and

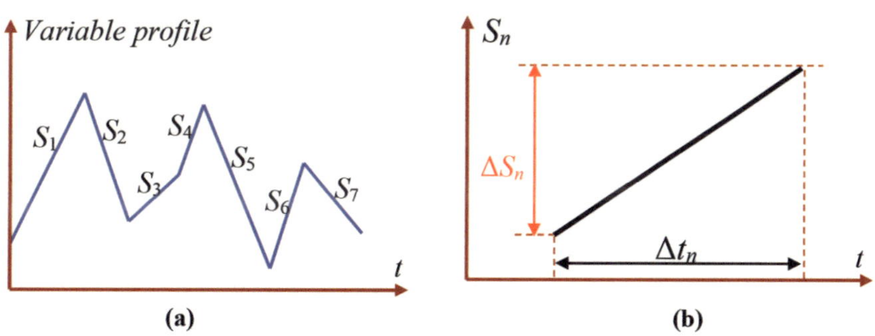

Figure 7.
Plot of battery cell number versus cycle duration. (a) Variable profile generated by segments, (b) Pattern parameters: ΔSn et Δtn.

they are extracted from simulation of the WT system over the reference profile of days.

The reference value of the storage useful energy E_{Sref} is given by the following equation:

$$E_{Sref} = \max E(t) - \min E(t) \quad (1)$$

with

$$E(t) = \int_0^t P_{BAT}(\tau)d\tau \quad t \in [0, \Delta t_{ref}] \quad (2)$$

To avoid oversizing during wide charge period (reduced consumption versus huge winds), the storage is only sized in discharge mode. Thus, E(t) is computed as a saturated integral, with 0 as upper limit. To take into account the reference wind cycle statistic features, an additional target indicator is considered: the cumulative distribution function $CDF(V_{ref})$ calculated from the corresponding probability density function PDF_{ref} which is evaluated on 20 equally spaced intervals between 0 and the maximum wind speed value V_{refmax} and related to the reference wind speed behavior.

Therefore, the inverse problem is set to minimize the global error ε in the synthesis profile process by

$$\varepsilon = \left(\frac{E_S - E_{Sref}}{E_{Sref}}\right)^2 + \left(\frac{P_{BAT\,MAX} - P_{BAT\,MAX\,ref}}{P_{BAT\,MAX\,ref}}\right)^2 + \left(\frac{P_{BAT\,MIN} - P_{BAT\,MIN\,ref}}{P_{BAT\,MIN\,ref}}\right)^2 + \varepsilon_{stat} \quad (3)$$

where the statistic error ε_{stat} denotes the mean squared error between both CDFs relative to reference and generated wind speed profiles:

$$\varepsilon_{stat} = \frac{1}{20} \times \sum_{k=1}^{20} \left(\frac{CDF(k) - CDF_{ref}(k)}{CDF_{ref}(k)}\right)^2 \quad (4)$$

All "ref" indexed variables are based on the reference wind profile of **Figure 8**. The inverse problem is solved with the clearing algorithm [17] using a population size of 100 individuals and a number of generations of 500,000.

Multiple optimization runs are performed with different compaction times $\Delta t_{compact}$. In order to guarantee a global error ε less than 10^{-}, the minimum compaction time was determined using dichotomous search. The values of ε versus compaction time are shown in **Table 3**. The minimum value for $\Delta t_{compact}$ assuring the completion of the target indicators with adequate accuracy is about 10 days. The generated wind profile is obtained from the aggregation of 109 elementary segments fulfilling all target indicators. The characteristics of this compact wind cycle and its CDF are displayed in **Figure 9**. It can be seen from this figure that the CDF of this wind profile closely coincides with that of the reference wind profile.

In **Table 4** a comparison between the target indicator values related to the storage sizing of the reference profile and the compact profile is generated with the clearing algorithm. A good agreement between those values indicates that the compact wind profile will lead to the same storage device sizing as with the reference wind profile on larger duration.

DOI: http://dx.doi.org/10.5772/intechopen.85613

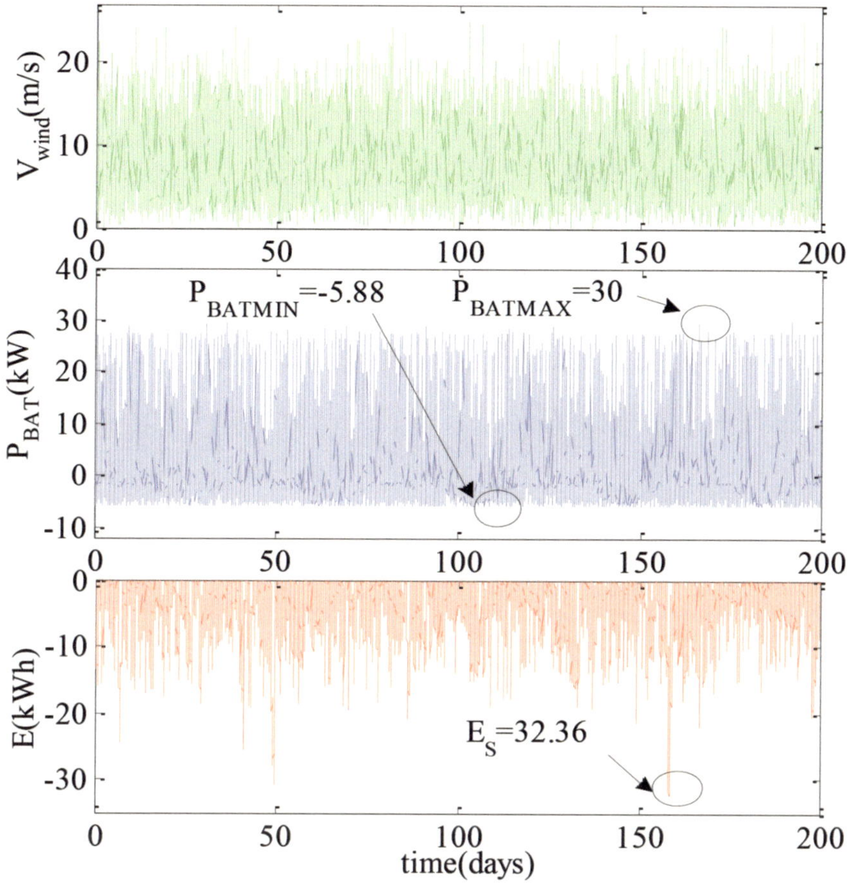

Figure 8.
Actual "reference" wind speed profile, storage power, and energy.

$\Delta t_{compact}$ (days)	40	20	10	5
Global error ε	≈8.10^{-3}	≈9.10^{-3}	≈9.10-3	≈7.10^{-2}

Table 3.
Influence of $\Delta T_{compact}$ on the global error ε.

4.3 Compact synthesis approach using wind-based targets

The selected target indicators are only related to the wind features: this approach can then be considered as generic in the case of any WT system whatever its sizing.

We first consider three indicators V_{max}, V_{min}, and $<V^3>$ representing the maximum and minimum speed values and the average cubic wind speed value. Note that $<V^3>$ is used instead of the average wind speed value $<V>$ because the WT power is directly proportional to the cubic wind speed value. Similarly to the previous approach, we also add the *CDF* as target indicator associated with the wind profile in order to take account of the wind statistic. Finally, we consider as last indicator related to "wind energy" with the variable E_V which is defined as

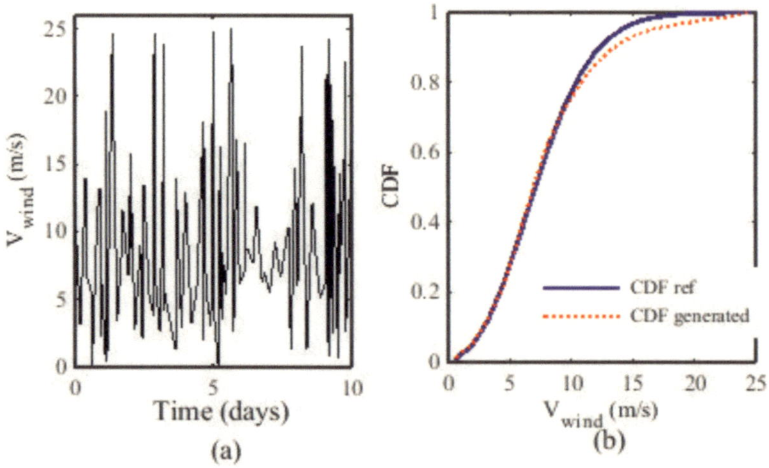

Figure 9.
Generated wind speed with corresponding CDF.

	Reference profile	Compact profile	Error (%)
P_{BATMAX} (kW)	30	30	0
P_{BATMIN} (kW)	≈-5.88	≈-5.82	0.1
ES (kWh)	32.36	32.4	0.12

Table 4.
Target indicators of the generated wind speed profile.

$$E_V = \max_{t\in[0,\Delta t]} E(t) - \min_{t\in[0,\Delta t]} E(t) \tag{5}$$

with

$$E(t) = \int_0^t \left(V^3(\tau) - <V^3>\right)d\tau \qquad t\in[0,\Delta t] \tag{6}$$

where E_V represents an "intermittent wind pseudo energy". In fact, E_V plays a similar role with ES in the previous approach for the storage system.

Note that the wind power being proportional to V^3, E_v is not actually an energy (in Joules or kWh) but can be seen as a "pseudo energy" which is qualitatively related to wind energy.

The global error ε to be minimized with this second approach can be expressed as

$$\varepsilon = \left(\frac{V_{\max} - V_{\max\, ref}}{V_{\max\, ref}}\right)^2 + \left(\frac{V_{\min} - V_{\min\, ref}}{V_{\min\, ref}}\right)^2 + \left(\frac{<V^3> - <V^3>_{ref}}{<V^3>_{ref}}\right)^2 + \left(\frac{E_V - E_{V\, ref}}{E_{V\, ref}}\right)^2 + \varepsilon_{stat} \tag{7}$$

where ε_{stat} is computed according to (3) and where the reference intermittent wind energy E_{Vref} is scaled according to the compact profile duration:

$$E_{V\ ref} = \frac{\Delta t_{compact}}{\Delta t_{ref}} \times E_{Vref}(\Delta t_{real}) \tag{8}$$

The inverse problem is solved with the clearing algorithm with the same control parameters as in the previous subsection. Multiple optimization runs were performed with different compaction times $\Delta t_{compact}$. The minimum value for this variable ensuring a global error less than 10^{-2} was identical to that found with the previous approach (i.e., 10 days). **Figure 10** shows the characteristics of the generated wind profile obtained for $\Delta t_{compact}$ = 10 days, from the aggregation of 130 elementary segments fulfilling all target indicators. The good agreement between the compact generated profile and the reference profile can also be observed in this figure in terms of CDF. Finally, **Table 5** shows that the values of the target indicators are very close in both cases.

Here, ε_{stat} is computed according to (3), and the reference intermittent wind energy E_{Vref} is scaled according to the compact profile duration:

$$E_{V\ ref} = \frac{\Delta t_{compact}}{\Delta t_{ref}} \times E_{Vref}(\Delta t_{real}) \tag{9}$$

The inverse problem is solved with the clearing algorithm with the same control parameters as in the previous subsection. Multiple optimization runs were performed with different compaction times $\Delta t_{compact}$. The minimum value for this variable ensuring a global error less than 10^{-2} was identical to that found with the previous approach (i.e., 10 days). **Figure 10** shows the characteristics of the generated wind profile obtained for $\Delta t_{compact}$ = 10 days, from the aggregation of 130 elementary segments fulfilling all target indicators. The good agreement between

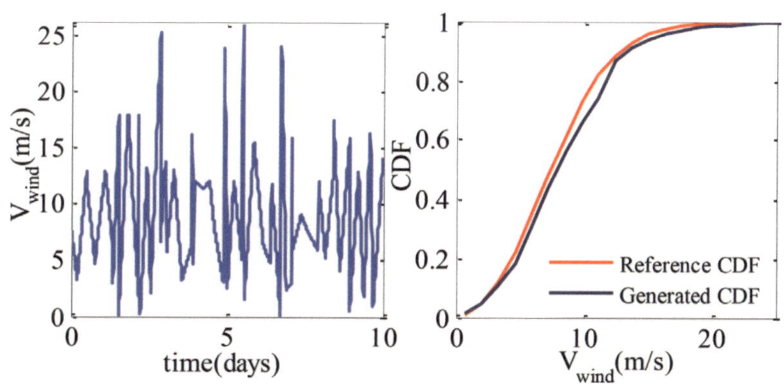

Figure 10.
Generated wind speed with corresponding CDF.

	Reference profile	Compact profile	Error (%)
V_{max} (m/s)	25.1	25.9	3.58
V_{min} (m/s)	0	0	0
$<V^3>$ (m^3/s^3)	876.4	871.4	0.57
E_V (m^3/s^2)	32.3	34.4	0.42

Table 5.
Target indicators of the reference versus generated wind speed profile with $\Delta t_{compact}$ = 10 days.

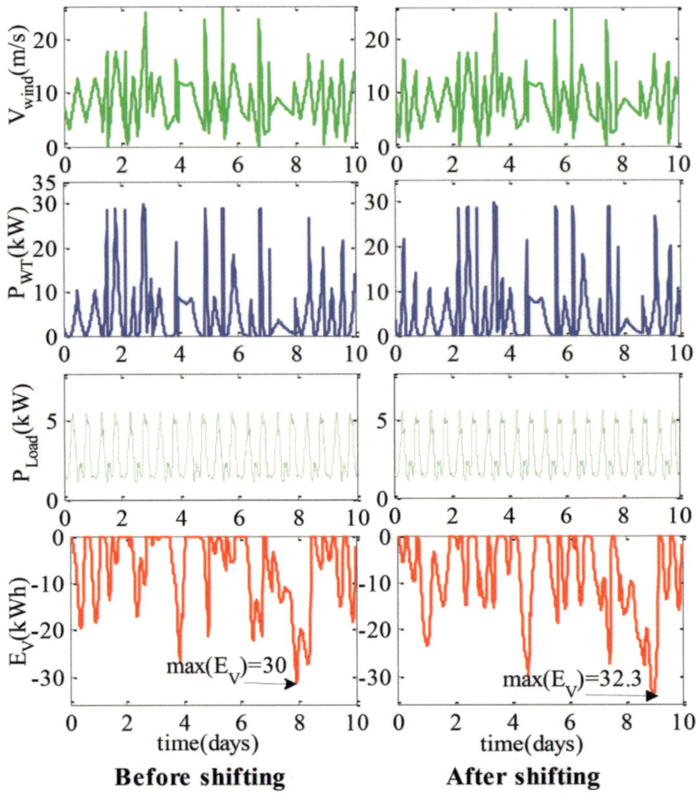

Figure 11.
Illustration of the phase shift of the wind profile (generated with the second method) on the battery sizing.

the compact generated profile and the reference profile can also be observed in this figure in terms of CDF. Finally, **Table 5** shows that the values of the target indicators are very close in both cases.

For comparison with the previous approach, we also give the sizing of the battery obtained from the simulation of the compact profile. It should be noted that contrarily to the first approach, the second one does not include phase correlations between wind and load profiles because it only considers wind speed variations to generate the compact wind speed profile. Consequently, the second approach does not ensure finding the most critical constraints on the storage device in terms of production—load phase shift. This can be a posteriori done by sequentially shifting the obtained wind profile on its 10-day time window in compliance with the deterministic load profile day to day repeated. The maximum storage energy quantity ES is computed for each phase shift and the highest (most critical) value is returned (see **Figure 11**). By this way, a value of 34.4 kWh is obtained for ES which is very close to that resulting from the reference profile simulation (i.e., 32.3 kWh).

5. Conclusions

In this chapter, new methodologies for sizing electrochemical devices into renewable energy systems are presented. As case of study, a battery bank devoted to a standalone WT system has been developed and compared. A passive WT structure, minimizing the number of sensors and the electronic part, has been chosen because of its reliability and its low cost. The two first sizing methodologies take account of stochastic features of wind energy potential in a particular location with a given deterministic power demand. These approaches are based on the exploitation of wind speed distribution from a Weibull law or directly the extracted power histogram at the WT output. It has been shown that a robust sizing of the storage device can be obtained from the stochastic generation of either the wind speed profile or the extracted WT output power using a specific algorithm. In this algorithm, the battery required active energy is calculated by upper saturated integration of the battery power. Two supplementary approaches have been developed for compacting wind speed profiles. These approaches consist in generating compact wind profiles by aggregating elementary-parameterized segments in order to fulfill target indicators representing the features of a reference wind profile of larger duration. The inverse problem involving the determination of the segment parameters is solved with an evolutionary algorithm. It is shown that both latter approaches are able to represent the main features of the reference profile in terms of wind farm potential and are also relevant for evaluating the critical conditions imposed to the battery storage (i.e., power and energy needs) in a hybrid WT system. All sizing methods have yielded roughly to same battery size but with different wind profiles durations. Statistical methods have provided a gain of 2.5 in time window reduction, while compact synthesis methods have led to a gain of 20. From these compacts profiles, subsequent reduction of the computation time should be obtained in the context of the optimization process of such systems. Note that this synthesis approach is very generic and could be extrapolated beyond the particular field of WT design and may be applied in the whole range of electrical engineering applications, by processing any types of environmental variables (wind speed but also temperature, sun irradiation, etc.).

Acknowledgements

This work was supported by the Tunisian Ministry of Higher Education, Research and Technology.

Author details

Malek Belouda

1 LAPER, Faculty of Sciences of Tunis, University of Tunis El Manar, Tunis, Tunisia

2 University of Carthage, The Higher Institute of Information Technologies and Communication ISTIC, Tunisia

*Address all correspondence to: malek.belouda@gmail.com

References

[1] Gavanidou E, Bakirtzis A. Design of a stand alone system with renewable energy sources using trade-off methods. IEEE Transactions on Energy Conversion. 1992;7(1)

[2] Chedid R, Rahman S. Unit sizing and control of hybrid wind-solar power systems. IEEE Transactions on Energy Conversion. 1997;**12**(1)

[3] Kellogg W, Nehrir M, Venkataramanan G, Gerez V. Generation unit sizing and cost analysis for stand-alone wind, photovoltaic, and hybrid wind/PV systems. IEEE Transactions On Energy Conversion. 1998;**13**(1)

[4] Bernard-Agustín JL, Dufo-Lopez R, Rivas-Ascaso DM. Design of isolated hybrid systems minimizing costs and pollutant emissions. Renewable Energy. 2006;**31**(14):2227-2244

[5] Senjyu T, Hayashi D, Yona A, Urasaki N, Funabashi T. Optimal configuration of power generating systems in isolated island with renewable energy. Renewable Energy. 2007;**32**:1917-1933

[6] Belfkira R, Nichita C, Reghem P, Barakat G. Modeling and optimal sizing of hybrid energy system. In: International Power Electronics and Motion Control Conference (EPE-PEMC); 1–3 September 2008; Poznan, Poland: IEEE PEMC

[7] Lim JH. Optimal combination and sizing of a new and renewable hybrid generation system. International Journal of Future Generation Communication and Networking. 2012;**5**(2)

[8] Tran DH, Sareni B, Roboam X, Espanet C. Integrated optimal design of a passive wind turbine system: An experimental validation. IEEE Transactions on Sustainable Energy. 2010;**1**(1):48-56

[9] Gupta SC, Kumar Y, Agnihotri G. REAST: Renewable energy analysis and sizing tool. Journal of Electrical Systems. 2011;**7**(2):206-224

[10] Protogeropoulos C, Brinkworth B, Marshall R. Sizing and techno-economical optimization for hybrid solar photovoltaic/wind power systems with battery storage. International Journal of Energy Research. 1997;**21**

[11] Morgan T, Marshall R, Brinkworth B. "ARES"—A refined simulation program for the sizing and optimization of autonomous hybrid energy systems. Solar Energy. 1997;**59**(4–6)

[12] Seeling-Hochmuth G. A combined optimization concept for the design and operation strategy of hybrid-PV energy systems. Solar Energy. 1997;**61**(2)

[13] Jaafar A, Sareni B, Roboam X. Signal synthesis by means of evolutionary algorithms. Journal on Inverse Problems in Science and Engineering. 2012; **20**(12):93-104

[14] Belouda M, Belhadj J, Sareni B, Roboam X. Battery sizing for a stand alone passive wind system using statistical techniques. In: 8th International Multi-Conference on Systems, Signals & Devices; Sousse, Tunisia. 2011

[15] Roboam X, Abdelli A, Sareni B. Optimization of a passive small wind turbine based on mixed Weibull-turbulence statistics of wind. In: Electrimacs 2008; Québec, Canada. 2008

[16] Schwefel H-P. Evolution and Optimum Seeking. Wiley; 1995

[17] Petrowski A. A clearing procedure as a niching method for genetic algorithms. In: Proceedings of the IEEE International Conference on Evolutionary Computation; Nagoya, Japan: 1996. pp. 798-803

Chapter 2

New Energy Management Concepts for Hybrid and Electric Powertrains: Considering the Impact of Lithium Battery and Ultracapacitor Aging

Francis Assadian, Kevin Mallon and Brian Walker

Abstract

During the lifetime of an energy storage system, its health deteriorates from use due to irreversible internal changes to the system. This degradation results in decreased capacity and efficiency of the battery or capacitor. This chapter reviews empirical aging models for lithium-ion battery and ultracapacitor energy storage systems. It will explore how operating conditions like large currents, high temperature, or deep discharge cycles impact the health of the energy storage system. After reviewing aging models, this chapter will then show how these models can be used in vehicle energy management control systems to reduce energy storage system aging. This includes both aging-aware control and control of hybrid energy storage systems (systems that include both a battery and an ultracapacitor).

Keywords: electric vehicle, hybrid vehicle, energy management, lithium ion, ultracapacitor, battery aging

1. Introduction

The internal combustion engine is a major contributor to greenhouse gas emissions and hydrocarbon pollution across the globe. Motor vehicles account for a major portion of pollutants such as carbon monoxide, nitrogen oxide, and volatile organic compounds [1]. Alternative powertrain vehicles (APVs), such as electric vehicles (EVs) and hybrid-electric vehicles (HEVs), are potential technological solutions to reduce transportation-sector emissions and fuel consumption. However, APVs require large amounts of battery-stored energy, which can be cost and weight prohibitive [2]. Degradation of the battery further adds to the lifetime cost of an APV, and battery degradation rate has been shown to be inversely correlated with fuel economy [3, 4]. Technologies that improve battery lifespan and fuel economy will reduce this lifetime cost and hasten the adoption of sustainable transportation.

Lithium-based batteries serve as the current main battery of choice for vehicle transportation because of their high energy density and ability for high cycle life. Improving cycle life of lithium batteries means limiting large currents in and out of the battery as much as possible to lower degradation and heat affects. Ultracapacitors (UC) can be added to vehicles to improve battery life by taking excess power away from the battery and storing it in temporary energy storage [5]. Capacitors can quickly unload power back into the system for high load situations such as a hard acceleration, taking away the need for a high-power drain from the battery.

This chapter will begin with a brief review of existing literature on empirical modeling of lithium-ion battery and ultracapacitor degradation. Then, a few select aging models will be reoriented for use in an APV energy management system (EMS). Finally, an example showing how to utilize these control-oriented models will be shown.

2. Energy storage aging review

2.1 Lithium ion battery aging

Aging of batteries is primarily caused by the formation of substrates in the chemical reaction pathways and the formation of cracks in the electrode materials from repeated stress cycles [6]. These aging mechanisms are accelerated by high charge and discharge rates, extreme battery temperatures, and deep depths of discharge [7]. Aging of the battery causes capacity fade (a decrease in the charge storage capacity) and power fade (a decrease in the battery efficiency). However, models of the cell chemistry that include the thermal and stress/strain relationships used to describe aging are computationally intensive and are ill-suited for use in APV EMSs [6, 8].

Research of battery aging in APVs instead tends to utilize empirical models [4, 9–14]. Using empirical aging models for vehicle battery degradation analysis provides a good trade-off between precision and complexity. These empirical models do not consider the physical or chemical processes of the battery degradation but instead approximate the battery's health by fitting experimental data to aging factors like charge throughput, calendar life, and number of charge/discharge cycles.

For instance, Refs. [9, 10, 15, 16] develop aging models that relate charge throughput to degradation, with temperature and current magnitude as additional stress factors. Refs. [17, 18] include depth of discharge as an additional stress factor, while [18] also distinguishes the impact of charging and discharging currents on battery degradation. The aging models for hybrid vehicle applications in [13, 14] consider a number of charge/discharge cycles and calendar life and use temperature, depth of discharge, and average state of charge as aging stress factors. Other models in the literature such as [8, 19, 20] use simple cycle counting to measure the state of health.

Current research works to integrate battery aging dynamics into these EMSs to form controllers that actively regulate battery degradation. In Ref. [4], the authors developed an SDP-based EMS for a parallel-HEV passenger vehicle that accounted for battery wear by mapping operating conditions to substrate growth, and associating substrate growth with battery state of health. The authors also analyzed how reducing battery aging increased the fuel consumption. In Refs. [4, 21], the authors developed a deterministic EMS for a parallel-HEV passenger vehicle that regulates battery degradation using a "severity factor" map: the control policy penalizes battery usage by an amount related to the severity of the operating conditions (in terms of temperature and current magnitude). The authors of [4] also showed an inverse correlation between the battery aging and fuel consumption.

2.2 Ultracapacitor aging

Lithium batteries have a high energy density but low power density, meaning that although they store large amounts of energy, that energy cannot be accessed quickly. Additionally, high currents to and from the battery are a stress factor for battery degradation. A potential solution to these problems is to integrate UCs into the energy storage system. UCs store energy in the electric field of an electrochemical double layer and have a high power density but low energy, allowing them to serve as complements to battery energy storage [5]. By integrating UCs into the powertrain, it becomes possible to meet the vehicle power requirements with a smaller battery and reduce battery degradation by restricting the magnitude of the current going to or from the battery [5, 22]. Aging of UCs is primarily dependent on time, temperature, and cell voltage [23–25].

Current research is interested in optimal control and sizing of the UC to reduce battery aging [26], and in particular how battery aging and fuel economy are jointly impacted. Some related work includes Ref. [27], in which the authors develop an optimal control policy to govern UC behavior and demonstrate clear aging improvements over a passive (uncontrolled) system. Refs. [28, 29] carried out a parametric study on battery degradation versus UC size in EVs, using a rule-based control system to govern power allocation. Ref. [30] developed a control strategy integrating UCs with lead-acid batteries in a HEV for battery life extension, and found that a 50% increase in battery cycle life would be needed for the UC to be cost-effective. Ref. [31] experimentally demonstrated a decrease in battery power fade and temperature rise in lithium-ion batteries due to UCs on an EV load profile.

3. Control-oriented aging

Meeting this goal of mitigating energy storage system degradation in APVs through control requires forming simplified models of the battery aging dynamics that can be included in or be used to generate an EMS. This chapter will summarize several approaches in the literature for using energy storage aging models for control applications.

3.1 Power-law model

Ref. [15] developed a lithium-ion battery empirical aging model for normalized battery capacity loss Q_{loss}, based on an Arrhenius equation. The model uses experimental data to relate battery degradation to on charge throughput Ah (in ampere-hours), current I_b (in C), and temperature T (in K).

$$Q_{loss} = A(I_b)\exp\left(\frac{-E_a + B|I_b|}{RT}\right)Ah^z \tag{1}$$

E_a, B, and z are fitted parameters, while $A(I_b)$ is a fitted function of current. R is the ideal gas constant. Here, $Q_{loss} = 0$ indicates a new battery, while $Q_{loss} = 0.2$, for example, indicates a 20% decrease in the storage capacity. This model treats current and temperature as static values. So, Eq. (1) can be differentiated to form a dynamic aging model as follows:

$$\frac{dQ_{loss}}{dt} = A(I_b)\exp\left(\frac{-E_a + B|I_b|}{RT}\right)\cdot z(Ah)^{z-1}\frac{dAh}{dt} \tag{2}$$

This model of aging has been used for aging control in, for example, [3, 4, 12].

3.2 Cycle life model

The Palmgren-Miner (PM) rule is a common method for analyzing fatigue life in mechanical systems and has been shown to effectively approximate the battery health over nonuniform charge and discharge cycles [8, 32, 33]. As per the PM rule, each charge and discharge cycle is considered to damage the battery by an amount related to the cycle life at that cycle's depth of discharge, charge and discharge current, and temperature. Ref. [18], for instance, models the cycle life of a battery as a function of depth of discharge DoD, charging current I_c, discharging current I_d, and temperature T.

$$CL = f(DoD, I_c, I_d, T) \tag{3}$$

Assume there is a charge-discharge cycle k with operating conditions DoD_k, $I_{c,k}$, $I_{d,k}$, and T_k, and so the cycle life for these operating conditions is CL_k. Then, under the PM rule, the damage D_k from cycle k is assumed to be

$$D_k = 1/CL_k \tag{4}$$

For multiple charge and discharge cycles, the damage from each cycle can be added to find the total damage D_{tot}. For the total damage up to cycle k,

$$D_{tot}(k) = \sum_{i=1}^{k} D_i \tag{5}$$

Cumulative damage of zero denotes that the battery is unaged while cumulative damage of one means the battery has reached the end of its life. Typically, 20% capacity fade indicates a battery's end of life. So,

$$Q_{loss}(k) = 0.2 \cdot D_{tot}(k) \tag{6}$$

However, the above method does not readily lend itself to use in control; full charge and discharge cycles can take a long time to develop, and the EMS must act at a faster rate. One possibility is that the energy management system could consider how its decision would cause the damage of the current cycle to grow or lessen. For instance, consider a battery to be at operating conditions DoD_j, $I_{c,j}$, $I_{d,j}$, and T_j. The EMS then makes some decision such that the operating conditions become DoD_k, $I_{c,k}$, $I_{d,k}$, and T_k. Using Eqs. (3) and (4), the change in damage due to the EMS's decision can be calculated as

$$\Delta D = D_k - D_j = \frac{1}{f(DoD_k, I_{c,k}, I_{d,k}, T_k)} - \frac{1}{f\left(DoD_j, I_{c,j}, I_{d,j}, T_j\right)} \tag{7}$$

Then, Eq. (7) could be used in formulating an energy management strategy, such that the EMS would seek to minimize the additional damage caused by each decision it makes.

3.3 Ultracapacitor aging

Ref. [23] provides the following model for ultracapacitor aging, where SoA is the state of aging where 0 indicates start of life and 1 indicates end of life.

$$\frac{dSoA}{dt} = \frac{1}{T_{life}^{ref}} \cdot \exp\left(\ln(2)\,\frac{\theta_c - \theta_c^{ref}}{\theta_0}\right) \cdot \left(\exp\left(\ln(2)\,\frac{V - V^{ref}}{V_0}\right) + K\right) \tag{8}$$

where θ_c is the UC temperature, V is the UC voltage, and the remaining variables (T_{life}^{ref}, θ_c^{ref}, θ_0, V^{ref}, V_0, and K) are experimentally fitted parameters. This model is ready to be used for control as is. Ref. [23] defines the UC end-of-life condition as similar to batteries: when the capacitance of the UC has faded by 20%.

4. Case study: electric vehicle with hybrid energy storage

This section develops a model for a hybrid energy storage system electric vehicle (HESS-EV)—specifically, an electric bus that uses a lithium-ion battery pack for energy storage and an ultracapacitor pack for handling large power requests. This example study will be used to show how active control of aging factors can improve the lifespan of the energy storage system without compromising energy consumption. This system is depicted in **Figures 1** and **2**.

4.1 Vehicle dynamics

For this study, a backward-facing quasi-static vehicle model [34] is used to represent the vehicle dynamics. In this model, it is assumed that the driver accurately follows the velocity of a given drive cycle, eliminating the need for a driver model and allowing the time-history of the electrical load placed on the powertrain to be calculated in advance.

This vehicle model, illustrated in **Figure 3**, considers inertial forces, aerodynamic drag, and rolling resistance (note that road incline is neglected for this chapter). The drag force is given by

$$F_{drag} = \frac{1}{2}\rho A_f C_D v_v^2 \tag{9}$$

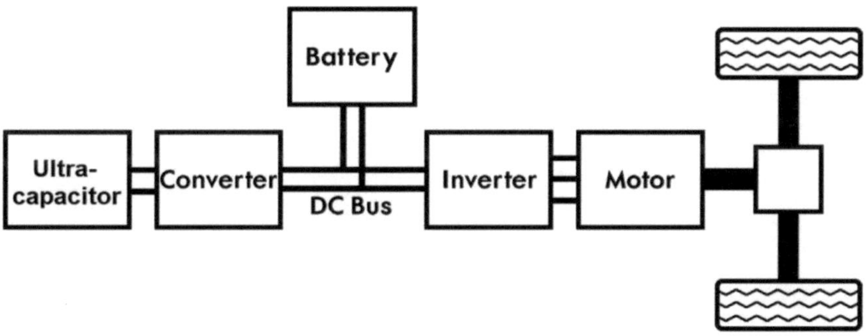

Figure 1.
HESS-EV model.

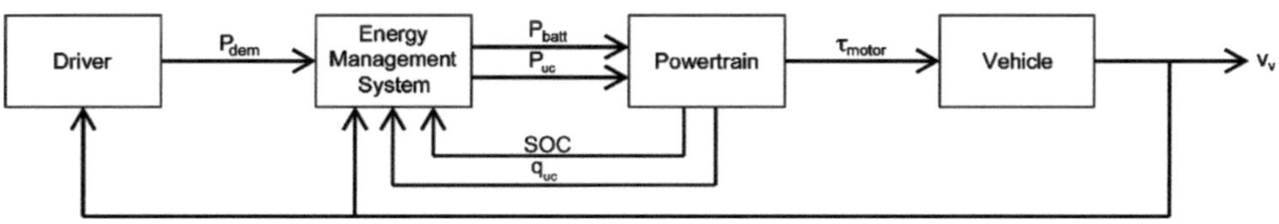

Figure 2.
HESS-EV block diagram.

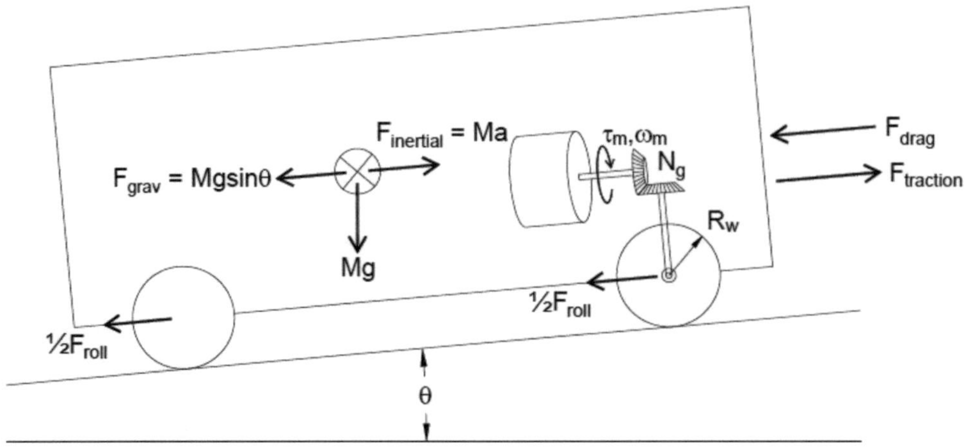

Figure 3.
Vehicle diagram.

where $\boldsymbol{\rho}$ is the air density, $\boldsymbol{A_f}$ is the frontal area, $\boldsymbol{C_D}$ is the drag coefficient, and $\boldsymbol{v_v}$ is the vehicle velocity. Rolling resistance is given by

$$\boldsymbol{F_{roll}} = \boldsymbol{M_v g C_R} \tag{10}$$

where M_v is the vehicle's total mass, g is the acceleration due to gravity, and C_R is the rolling resistance coefficient. In a backward-facing model, the inertial force is determined from the vehicle acceleration and the vehicle mass as

$$F_{inertial} = M_{eq}\frac{dv_v}{dt}. \tag{11}$$

M_{eq} is the mass of the bus plus the equivalent mass due to the rotational inertia of the motor and wheels.

$$M_{eq} = M_v + 4J_w\left(\frac{1}{R_w}\right)^2 + J_m\left(\frac{N_{fd}N_{gb}}{R_w}\right)^2, \tag{12}$$

where J_w is the rotational inertia of one wheel, J_m is the rotational inertia of the motor, R_w is the wheel radius, N_{fd} is the final drive ratio, and N_{gb} is the gearbox ratio. The acceleration term in Eq. (11) is approximated from a given velocity profile according to

$$\frac{dv_v}{dt}(t) \approx \frac{v_v(t+\Delta t) - v_{v(t-\Delta t)}}{2\Delta t}. \tag{13}$$

The inertial, drag, and rolling resistance forces sum together to give the traction force on the bus.

$$F_{traction} = F_{inertial} + F_{drag} + F_{roll} \tag{14}$$

Parameter values for the vehicle model can be found in **Table 1**. The bus is assumed to be fully loaded and at its maximum allowable weight. Vehicle parameters are estimated from existing literature on bus simulation [35–37].

Parameter	Variable	Value
Vehicle mass	M_v	18,181 kg
Frontal area	A_f	8.02 m^2
Drag coefficient	C_D	0.55
Roll resistance coefficient	C_R	0.008
Wheel inertia	Jw	20.52 kg-m^2
Motor inertia	Jm	0.277 kg-m^2
Wheel radius	R_w	0.48 m
Final drive ratio	Nfd	5.1:1
Gearbox ratio	N_{gb}	5:1

Table 1.
Vehicle physical parameters.

4.2 Powertrain model

This subsection describes the modeling of the HESS-EV powertrain, including the transmission, motor, battery, and ultracapacitor subsystems, as indicated in **Figure 1**. The goal of the vehicle model is to capture the primary forces on the vehicle while maintaining model simplicity. Both these make simulation of the system easier and make optimal control methods, such as dynamic programming or model-predictive control, less computationally complex. Otherwise, the energy management system might suffer from the "curse of dimensionality".

4.2.1 Transmission

Next, the vehicle speed and traction force are transformed into motor torque and motor speed. Assuming transmission efficiency of η_{trans}, represented as torque losses, the motor torque is given by

$$\tau_m = \begin{cases} \left(\frac{R_w}{N_{fd}N_{gb}} F_{traction}\right) / \eta_{trans} & F_{traction} \geq 0 \\ \left(\frac{R_w}{N_{fd}N_{gb}} F_{traction}\right) \cdot \eta_{trans} & F_{traction} < 0 \end{cases} \tag{15}$$

and the motor speed is given by

$$\omega_m = \frac{N_{fd}N_{gb}}{R_w} v_v \tag{16}$$

Then, the mechanical power needed to drive the vehicle P_{mech} can be expressed in terms of the motor torque and angular velocity.

$$P_{mech} = \tau_m \cdot \omega_m \tag{17}$$

Here, positive P_{mech} indicates acceleration. Parameter values for the transmission can be found in **Table 1**.

4.2.2 Motor and power electronics

The electrical power demand of the motor, P_{dem}, is calculated from P_{mech} and an efficiency parameter η_{motor}, $0 < \eta_{motor} < 1$. ηmotor is a function of τ_m and ω_m and is determined from a static efficiency map.

$$P_{dem} = \begin{cases} P_{mech}/\eta_{trans} & \tau_m \geq 0 \\ P_{mech} \cdot \eta_{trans} & \tau_m < 0 \end{cases} \tag{18}$$

The efficiency map is obtained from the National Renewable Energy Laboratory's Advanced Vehicle Simulator (ADVISOR) data library [38] and scaled to the appropriate size using the scaling method in [5]. It includes both the motor efficiency and the efficiency of the power electronics. The modeled vehicle utilizes a 250 kW AC induction motor.

The power demand for the electric motor is provided by battery power P_{batt} and ultracapacitor power P_{uc}. As part of the quasi-static simulation, it is assumed that the power demand is always met.

$$P_{dem} = P_{batt} + P_{uc} \tag{19}$$

Because P_{dem} is set by the drive cycle and P_{uc} is a controlled variable, P_{batt} is fixed and dependent on both P_{dem} and P_{uc}. Therefore, Eq. (19) can be rewritten as

$$P_{batt} = P_{dem} – P_{uc} \tag{20}$$

4.3 Energy storage systems

The previous subsections detailed how the driver's electrical power request would be determined. As depicted in **Figure 4**, the EMS decides how that power is split between the lithium-ion battery and the ultracapacitor. This subsection will first detail the modeling of the battery, followed by modeling of the ultracapacitor.

4.3.1 Battery

This HESS-EV uses lithium-ion batteries represented by the simple battery model shown in **Figure 5**, where V_{cell} is the open-circuit voltage (OCV) of a single battery cell, while R_{cell} represents the combined effects of ohmic resistances, diffusion resistances, and charge-transfer resistances [5]. This quasistatic model requires only a single state variable, state of charge (SOC). The OCV as well as the resistance are considered to vary with SOC per experimental data for a lithium-iron-phosphate battery [39].

The equivalent resistance of the complete battery pack is given by

$$R_{eq} = R_{cell} \frac{N_{ser}}{N_{par}} \tag{21}$$

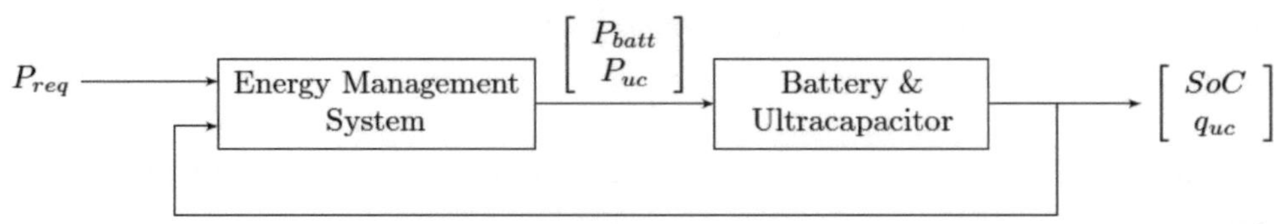

Figure 4.
Battery and UC block diagram.

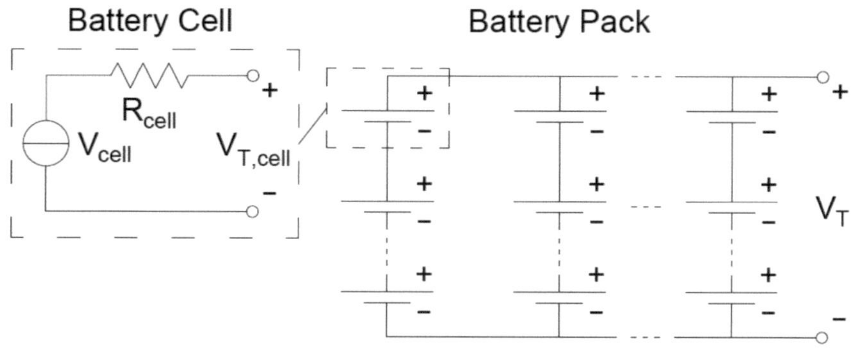

Figure 5.
Battery cell and battery pack equivalent circuit.

where N_{ser} is the number of cells in series, N_{par} is the number in parallel, and R_{cell} is the resistance of a single cell. The open circuit voltage of the vehicle battery pack is likewise given by

$$V_{ocv} = N_{ser} \cdot V_{cell} \tag{22}$$

Using the equivalent circuit in **Figure 4**, the battery terminal voltage can be found from the OCV and battery power P_{batt} as follows:

$$I_{batt} = P_{batt}/V_T \tag{23}$$

$$V_T = V_{ocv} - I_{batt} \cdot R_{eq} \tag{24}$$

Then, substituting Eq. (23) into Eq. (24) and solving yields

$$V_T^2 = V_{ocv} \cdot V_T - P_{batt}R_{eq} \tag{25}$$

$$V_T = \frac{1}{2}\left(V_{ocv} + \sqrt{V_{ocv}^2 - 4P_{batt}R_{eq}}\right) \tag{26}$$

V_T can then be substituted back into Eq. (23) to obtain the battery current, which can be integrated to obtain the state of charge.

$$SOC(k+1) = SOC(k) + \Delta t \cdot \frac{I_{batt}}{Q_{batt}}, \tag{27}$$

where Q_{batt} is the capacity of the battery pack in coulombs and Δt is the timestep.

The parameters for the battery model can be found in **Table 2**. The number of cells in series was chosen so that the OCV would be in line with the

Parameter	Variable	Value
Battery cells in parallel	*Npar*	400 cells
Parallel sets in series	*Nser*	100 sets
Total charge capacity	*Qbatt*	340 Ah
Battery temperature	*T*	35°C.

Table 2.
Battery parameters.

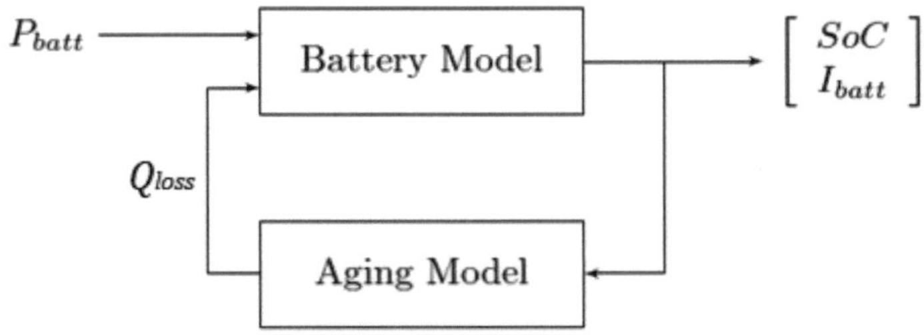

Figure 6.
Battery aging and dynamics block diagram.

recommendations in [40]. The number of cells in parallel was chosen so that the bus can be driven for 4 hours continuously to meet the power requirements of [40, 41].

For this example, the cycle counting method described in Eqs. (3)–(7) is used, with the aging model in Ref. [18] used to determine cycle life. This is illustrated in **Figure 6**. The battery is assumed to operate at a constant 35°C.

4.3.2 Ultracapacitor

The ultracapacitor model is similar in nature to the battery model, so the dynamics here will be presented more briefly. For this study, a second-order equivalent circuit based on the 100F ultracapacitor model in [42] is used to model the individual ultracapacitors. Parameters for this model are given in **Table 3**. Like with the battery, the ultracapacitor pack consists of ultracapacitors arranged in series and parallel as shown in **Figure 7**.

As shown in **Figure 1**, the UC is connected to the DC bus through a converter, so that the voltage of the UC pack is independent of the voltage at the DC bus. The ultracapacitor pack takes on total power P_{uc} and has N_{pc} cells in parallel per set and N_{sc} sets of cells in series. Then, the power going to each individual cell is

$$P_{uc,cell} = \frac{P_{uc}}{N_{pc} N_{sc}} \tag{28}$$

Let q_1 be the charge that corresponds to the C_1 capacitor, and let q_2 be the charge that corresponds to the C_2 capacitor. For a given power $P_{uc,cell}$, the current and terminal voltage can be found in a similar manner to Eqs. (23)–(26).

$$I_{uc} = P_{uc,cell}/V_{T,uc} \tag{29}$$

$$V_{T,uc} = q_2/C_2 - I_{uc} R_2 \tag{30}$$

Parameter	Variable	Value
UC parallel cells	*Npc*	100
UC series sets	*Nsc*	100
Resistor 1	*R1*	29.6 mΩ
Capacitor 1	*C1*	31.7 F
Resistor 2	*R2*	14.7 mΩ
Capacitor 2	*C2*	74.1781 F
Temperature	θ	45°C

Table 3.
Ultracapacitor parameters.

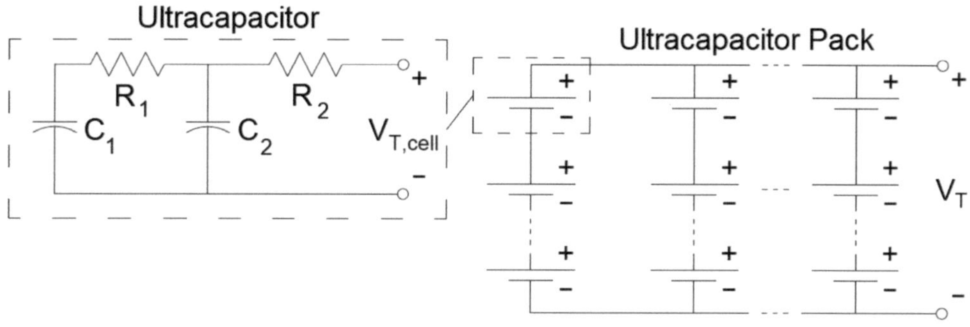

Figure 7.
Ultracapacitor pack equivalent circuit.

Then, substituting Eq. (23) into Eq. (24) and solving yields

$$V_{T,uc}^2 = q_2/C_2 \cdot V_{T,uc} - P_{batt}R_{eq} \tag{31}$$

$$V_{T,uc} = \frac{1}{2}\left(q_2/C_2 + \sqrt{q_2^2/C_2^2 - 4P_{uc,cell}R_2}\right) \tag{32}$$

V_T can then be substituted back into Eq. (23) to obtain the battery current. Then, the state equations for the two capacitors are

$$\dot{q}_1 = 1/R_1\left(q_1/C_1 - q_2/C_2\right) \tag{33}$$

$$\dot{q}_2 = I_{uc} - 1/R_1\left(q_1/C_1 - q_2/C_2\right) \tag{34}$$

The total charge in the capacitor q_{uc} is given by

$$q_{uc} = q_1 + q_2 \tag{35}$$

For this example, the cycle counting model in [23] and Section 3.3 is used. This is illustrated in **Figure 8**. The ultracapacitor is assumed to operate at a constant 45°C.

4.4 Energy management system

A deterministic dynamic programming (DDP) controller is used for the EMS. DDP is a form of optimal control—in this example, the DDP controller solves the following optimization problem over a known velocity profile that is N steps long. For this system, the minimum, target, and maximum charge are $q_{min} = 50\ C$, $q_{tgt} = 150\ C$, and $q_{max} = 250\ C$, and the minimum and maximum UC power are $P_{uc,\min} = -62.5W$ and $P_{uc,\max} = 62.5W$ so as not to exceed manufacturer specified operating conditions [42]. Note that these are constraints on each cell, not the entire pack. The battery power constraints will be discussed shortly.

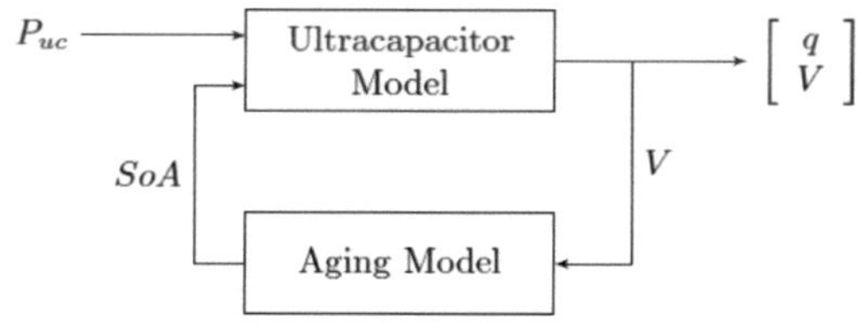

Figure 8.
Ultracapacitor aging & dynamics block diagram.

$$\begin{aligned} &\text{minimize} \sum_{i=0}^{N} \left(q_{uc}(i) - q_{uc,\,tgt}\right)^2 + Q_1 \cdot (\Delta D(i))^2 \\ &\textit{subject to } q_{min} < q_{uc} < q_{max} \\ &\qquad\qquad P_{uc,\,\min} < P_{uc} < P_{uc,\,\max} \\ &\qquad\qquad P_{batt,\,\min} < P_{batt} < P_{batt,\,\max} \end{aligned} \quad (36)$$

In plain terms, the DDP controller finds how to split power between the battery and ultracapacitor in such a way as to

1. Keep the UC charge near a target value

2. Minimize the aging of the battery

3. Ensure that the UC charge, UC power, and battery power stays within given bounds

The method to solve DDP problems can be found in Ref. [5].

In order to demonstrate the benefit of actively controlling aging, two versions of the controller will be tested:

1. Load-leveling DDP: Q_1 is set to zero. A battery power constraint of $P_{batt,\,min} = -3.2W$ and $P_{batt,\,max} = 3.2W$ per cell prevents large power (and therefore large current) going to the battery, and the cost function will bring the UC charge back to the target afterwards.

2. Active Aging Control: Battery power is unconstrained, but battery damage is directly penalized. A range of values are used for Q_1.

4.5 Simulation

The HESS-EV is simulated on the Manhattan Bus Cycle [43]—an urban bus velocity profile—for 4 hours at a time. After each simulation, the aging for the battery and ultracapacitor is measured. The battery capacity and ultracapacitor capacitances are then updated, and the next simulation begins. This process is repeated until the battery reaches its end of life.

4.6 Results

The lifespan of the battery in years is estimated by measuring the total kilometers driven before the battery reached the end of its life, and then using the Federal Highway Administration's average annual kilometers driven by transit busses [44] to convert the kilometers driven into an approximate number of years. Additionally, the ultracapacitor degradation and average kWh/km for the HESS-EV over the lifespan of the battery are measured.

Figure 9 shows a comparison between battery lifespan and fuel economy for both controller types. Clearly, the aging-aware control outperforms the load-leveling type controller: In all cases, the battery with actively-controlled aging has a longer lifespan.

Additionally, this figure shows a trade-off between efficiency and battery lifespan: As the battery lifespan increases with greater penalties on battery damage, the energy efficiency of the vehicle drops. This is because greater penalties on aging cause more current to pass through UC in order to reduce the load on the battery. In

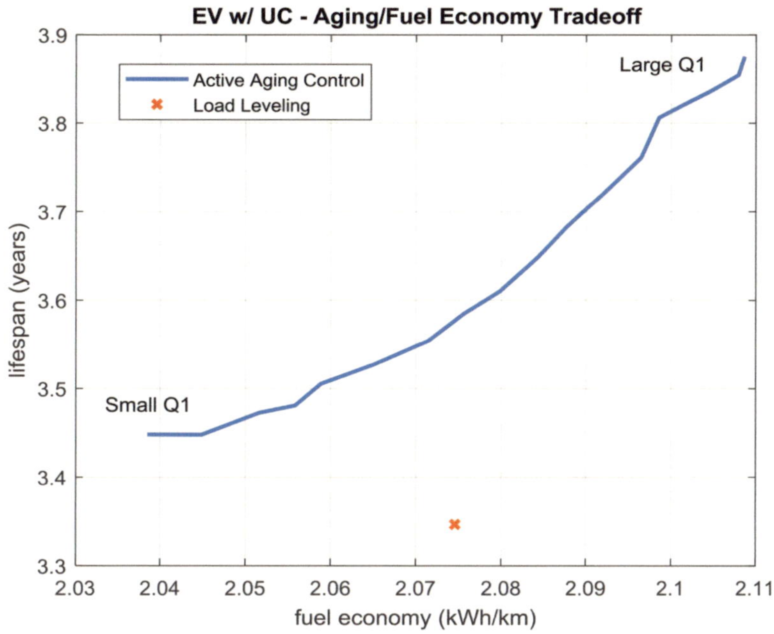

Figure 9.
Lifespan vs. fuel economy for HESS-EV, where Q_1 is the penalty on battery damage.

turn, there are more power losses due to the internal resistances of the capacitor pack. This is illustrated in **Figure 10**, which shows the charge in the UC for two different values of Q_1. One can see how a small penalty on Q_1 means the controller will focus mostly on keeping the UC charge near the target value; this in turn means less current through the UC, so less power losses from the UCs.

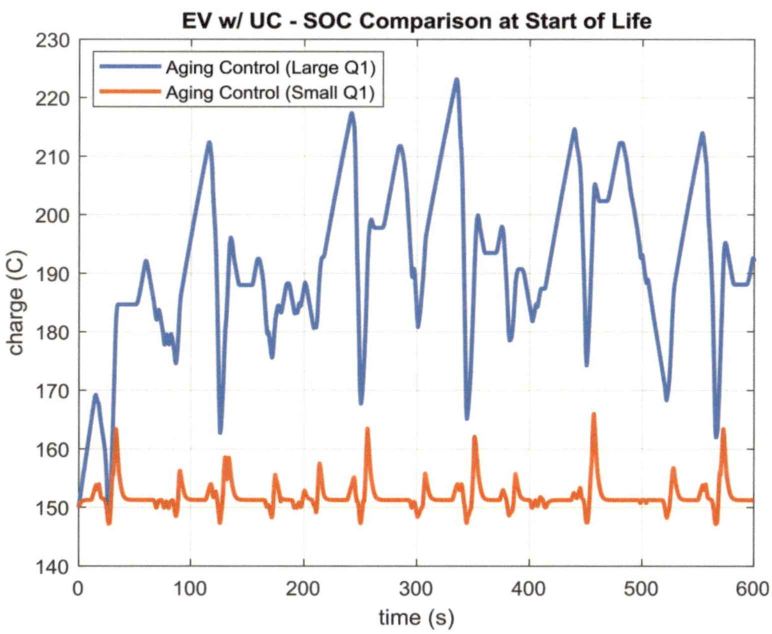

Figure 10.
Ultracapacitor charge for two different values of Q_1.

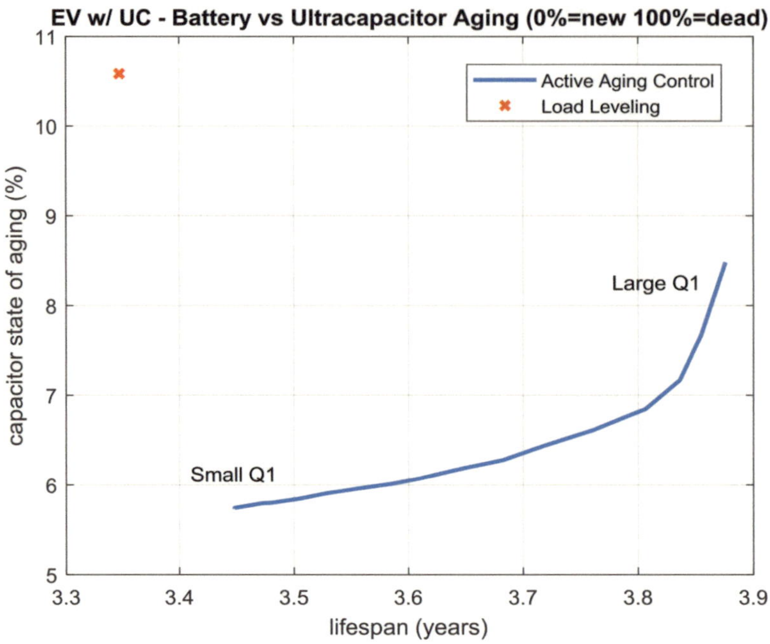

Figure 11.
Battery lifespan vs. UC aging. Q_1 is the penalty on battery damage.

Figure 11 shows the relationship between battery degradation and ultracapacitor degradation. Two things are apparent: one, there is an inverse relationship between the two—increasing battery lifespan comes at the cost of reducing ultracapacitor lifespan. A cause of this can be observed in **Figure 10**. Second, ultracapacitor degradation happens much more slowly than battery degradation. Despite the battery reaching the end of its lifespan, the ultracapacitor ages no more than 6–11%.

5. Conclusions

Deterioration of energy storage systems is inevitable, but by understanding the process it becomes possible to control and slow the capacity and efficiency fade. This chapter covered empirical aging models for lithium-ion and ultracapacitor systems and their use in vehicle energy management. First, existing work on different lithium ion and ultracapacitor aging models was reviewed, as well as those models' application in energy management control strategies. After reviewing aging models and discussing how to adapt empirical aging models for control, a case study was carried out on an ultracapacitor-augmented electric vehicle to show how actively controlling aging can improve an EMS. This case study included the steps necessary to model the vehicle and powertrain dynamics as well as simple or quasistatic models of the battery and ultracapacitor. DDP was generally used in two types of controllers: a load-leveling type controller that was unaware of aging dynamics, and a "smart" controller that incorporated battery aging dynamics into its design. When simulated, the aging-aware controller outperformed the simple controller, offering longer battery lifespan without any cost in fuel economy or vehicle performance. This demonstrates how advanced control—making EMSs aware of energy storage aging dynamics—can improve the efficiency and viability of alternative powertrain vehicles.

Acknowledgements

This work was supported by the University of California, Davis and by the Vertically Integrated Projects (VIP) program at UC Davis.

Conflict of interest

The authors declare no conflict of interest.

Author details

Francis Assadian*, Kevin Mallon and Brian Walker
University of California—Davis, Davis, CA, USA

*Address all correspondence to: fassadian@ucdavis.edu

References

[1] Abdel-Rahman A. On the emissions from internal-combustion engines: A review. International Journal of Energy Research. 1998;**22**:483-513

[2] Lee T-K, Filipi Z. Impact of model-based lithium-ion battery control strategy on battery sizing and fuel economy in heavy-Duty HEVs. SAE International Journal of Commercial Vehicles. 2011;**4**:198-209

[3] Moura SJ, Stein JL, Fathy HK. Battery-health conscious power management in plug-in hybrid electric vehicles via electrochemical modeling and stochastic control. IEEE Transactions on Control Systems Technology. 2013;**21**:679-694

[4] Suri G, Onori S. A control-oriented cycle-life model for hybrid electric vehicle lithium-ion batteries. Energy. 2016;**96**:644-653

[5] Guzzella L, Sciarretta A. Vehicle Propulsion Systems. 3rd ed. Berlin, Heidelberg: Springer Berlin Heidelberg; 2013. DOI: 10.1007/978-3-642-35913-2

[6] Ramadesigan V, Northrop PWC, De S, Santhanagopalan S, Braatz RD, Subramanian VR. Modeling and simulation of lithium-ion batteries from a systems engineering perspective. Journal of the Electrochemical Society. 2012;**159**:R31-R45

[7] Pelletier S, Jabali O, Laporte G, Veneroni M. Battery degradation and behaviour for electric vehicles: Review and numerical analyses of several models. Transportation Research Part B: Methodological. 2017;**103**:163-164

[8] Safari M, Morcrette M, Teyssot A, Delacourt C. Life-prediction methods for lithium-ion batteries derived from a fatigue approach I. Introduction: Capacity-loss prediction based on damage accumulation. Journal of the Electrochemical Society. 2010;**157**: A713-A720

[9] Petit M, Prada E, Sauvant-Moynot V. Development of an empirical aging model for Li-ion batteries and application to assess the impact of vehicle-to-grid strategies on battery lifetime. Applied Energy. 2016;**172**: 398-407

[10] Bishop JDK, Axon CJ, Bonilla D, Tran M, Banister D, McCulloch MD. Evaluating the impact of V2g services on the degradation of batteries in PHEV and EV. Applied Energy. 2013;**111**: 206-218

[11] Cordoba-Arenas A, Onori S, Rizzoni G. A control-oriented lithium-ion battery pack model for plug-in hybrid electric vehicle cycle-life studies and system design with consideration of health management. Journal of Power Sources. 2015;**279**:791-808

[12] Onori S, Spagnol P, Marano V, Guezennec Y, Rizzoni G. A new life estimation method for lithium-ion batteries in plug-in hybrid electric vehicles applications. International Journal of Power Electronics. 2012;**4**: 302-319

[13] Millner A. Modeling lithium ion battery degradation in electric vehicles. In: 2010 IEEE Conference on Innovative Technologies for an Efficient and Reliable Electricity Supply; IEEE. 2010. pp. 349-356

[14] Guenther SB, Hennings W, Waldowski P, Danzer MA. Model-based investigation of electric vehicle battery aging by means of vehicle-to-grid scenario simulations. Journal of Power Sources. 2013;**239**:604-610

[15] Wang J, Liu P, Hicks-Garner J, Sherman E, Soukiazian S, Verbrugge M, et al. Cycle-life model for

graphite-LiFePO4 cells. Journal of Power Sources. 2011;**196**:3942-3948

[16] Spotnitz R. Simulation of capacity fade in lithium-ion batteries. Journal of Power Sources. 2003;**113**:72-80

[17] Sarasketa-Zabala E, Martinez-Laserna E, Berecibar M, Gandiaga I, Rodriguez-Martinez LM, Villarreal I. Realistic lifetime prediction approach for Li-ion batteries. Applied Energy. 2016;**162**:839-852

[18] Omar N, Monem MA, Firouz Y, Salminen J, Smekens J, Hegazy O, et al. Lithium iron phosphate based battery—Assessment of the aging parameters and development of cycle life model. Applied Energy. 2014;**113**:1575-1585

[19] DiOrio N, Dobos A, Janzou S, Nelson A, Lundstrom B. Technoeconomic Modeling of Battery Energy Storage in SAM. tech. rep., National Renewable Energy Laboratory (NREL) NREL/TP-6A20–64641; 2015

[20] Marano V, Onori S, Guezennec Y, Rizzoni G, Madella N. Lithium-ion batteries life estimation for plug-in hybrid electric vehicles. In: 2009 IEEE Vehicle Power and Propulsion Conference; Sept. 2009. pp. 536-543

[21] Cordoba-Arenas A, Onori S, Guezennec Y, Rizzoni G. Capacity and power fade cycle-life model for plug-in hybrid electric vehicle lithium-ion battery cells containing blended spinel and layered-oxide positive electrodes. Journal of Power Sources. 2015;**278**: 473-483

[22] Williamson SS. Energy Management Strategies for Electric and Plug-in Hybrid Electric Vehicles. New York, NY: Springer New York; 2013. DOI: 10.1007/978-1-4614-7711-2

[23] Kovaltchouk T, Multon B, Ahmed HB, Aubry J, Venet P. Enhanced aging model for supercapacitors taking into account power cycling: Application to the sizing of an energy storage system in a direct wave energy converter. IEEE Transactions on Industry Applications. 2015;**51**:2405-2414

[24] Kovaltchouk T, Ahmed HB, Multon B, Aubry J, Venet P. An aging-aware life cycle cost comparison between supercapacitors and Li-ion batteries to smooth Direct Wave Energy Converter production. In: 2015 IEEE Eindhoven PowerTech; June 2015. pp. 1-6

[25] Hammar A, Venet P, Lallemand R, Coquery G, Rojat G. Study of accelerated aging of supercapacitors for transport applications. IEEE Transactions on Industrial Electronics. 2010;**57**:3972-3979

[26] Miller JM, Sartorelli G. Battery and ultracapacitor combinations #x2014; Where should the converter go? In: 2010 IEEE Vehicle Power and Propulsion Conference; 2010. pp. 1–7

[27] Song Z, Li J, Han X, Xu L, Lu L, Ouyang M, et al. Multi-objective optimization of a semi-active battery/supercapacitor energy storage system for electric vehicles. Applied Energy. 2014;**135**:212-224

[28] Shen J, Dusmez S, Khaligh A. Optimization of sizing and battery cycle life in battery/ultracapacitor hybrid energy storage systems for electric vehicle applications. IEEE Transactions on Industrial Informatics. 2014;**10**: 2112-2121

[29] Akar A, Tavlasoglu Y, Vural B. An energy management strategy for a concept battery/ultracapacitor electric vehicle with improved battery life. IEEE Transactions on Transportation Electrification. 2017;**3**:191-200

[30] Carter R, Cruden A, Hall PJ. Optimizing for efficiency or battery life in a battery/supercapacitor electric

vehicle. IEEE Transactions on Vehicular Technology. 2012;**61**:1526-1533

[31] Zhao C, Yin H, Ma C. Quantitative evaluation of LiFePO battery cycle life improvement using ultracapacitors. IEEE Transactions on Power Electronics. 2016;**31**:3989-3993

[32] Zhou C, Qian K, Allan M, Zhou W. Modeling of the cost of EV battery wear due to V2G Application 621 in power systems. IEEE Transactions on Energy Conversion. 2011;**26**:1041-1050

[33] Mallon KR, Assadian F, Fu B. Analysis of on-board photovoltaics for a battery electric bus and their impact on battery lifespan. Energies. 2017;**10**:943

[34] Mohan G, Assadian F, Longo S. Comparative analysis of forward-facing models vs backwardfacing models in powertrain component sizing. In: IET Hybrid and Electric Vehicles Conference 2013 (HEVC 2013); 2013. pp. 1-6

[35] Zeng X, Yang N, Wang J, Song D, Zhang N, Shang M, et al. Predictive-model-based dynamic coordination control strategy for power-split hybrid electric bus. Mechanical Systems and Signal Processing. Aug. 2015;**6061**: 785-798

[36] Sangtarash F, Esfahanian V, Nehzati H, Haddadi S, Bavanpour MA, Haghpanah B. Effect of different regenerative braking strategies on braking performance and fuel economy in a hybrid electric bus employing CRUISE vehicle simulation. SAE International Journal of Fuels and Lubricants. 2008;**1**(1):828-837

[37] Wang BH, Luo YG, Zhang JW. Simulation of city bus performance based on actual urban driving cycle in China. International Journal of Automotive Technology. 2008;**9**(4): 501-507

[38] Markel T, Brooker A, Hendricks T, Johnson V, Kelly K, Kramer B, et al. ADVISOR: A systems analysis tool for advanced vehicle modeling. Journal of Power Sources. 2002;**110**(2):255-266

[39] Erdinc O, Vural B, Uzunoglu M. A dynamic lithium-ion battery model considering the effects of temperature and capacity fading. In: 2009 International Conference on Clean Electrical Power, ICCEP; 2009. pp. 383-386. DOI: 10.1109/ICCEP.2009.5212025

[40] Nelson RF. Power requirements for batteries in hybrid electric vehicles. Journal of Power Sources. 2000;**91**(1): 2-26

[41] Fauvel C, Napal V, Rousseau A. Medium and heavy duty hybrid electric vehicle sizing to maximize fuel consumption displacement on real world drive cycles. Los Angeles, California: Power (W); 2012. pp. 0-22. Available from: https://www.autonomie.net/docs/5%20-% 20Presentations/Heavy%20duty/CF%20-%20EVS26.pdf

[42] Dougal RA, Gao L, Liu S. Ultracapacitor model with automatic order selection and capacity scaling for dynamic system simulation. Journal of Power Sources. 2004;**126**:250-257

[43] Barlow TJ, Latham S, Mccrae IS, Boulter PG. A Reference Book of Driving Cycles for Use in the Measurement of Road Vehicle Emissions. TRL Published Project Report; 2009. Available from: https://trid.trb.org/view/909274

[44] FHWA (Federal Highway Administration). Highway Statistics 2013; 2013

Section 2

Control of Energy Storage Systems

Chapter 3

Control Mechanisms of Energy Storage Devices

Mahmoud Elsisi

Abstract

The fast acting due to the salient features of energy storage systems leads to using of it in the control applications in power system. The energy storage systems such as superconducting magnetic energy storage (SMES), capacitive energy storage (CES), and the battery of plug-in hybrid electric vehicle (PHEV) can storage the energy and contribute the active power and reactive power with the power system to extinguish the rapid change in load demands and the renewable energy sources (RES). This chapter gives an overview about the modeling of energy storage devices and methods of control in them to adjust steady outputs.

Keywords: energy storage devices, superconducting magnetic energy storage (SMES), capacitive energy storage (CES), plug-in hybrid electric vehicle (PHEV)

1. Introduction

With the increasing of distributed generator (DG) technologies, large numbers of DGs are connected with the grid in different forms, such as wind and solar power systems [1–3]. Because of the fluctuations of their output power, energy storage devices are utilized to adjust steady outputs [4, 5]. In fact, the characteristics of the different storage devices vary widely, including the amount of energy stored and the time for which this stored energy is required to be retained or released [6, 7]. The superconducting magnetic energy storage (SMES), superconducting capacitive energy storage (CES), and the battery of plug-in hybrid electric vehicle (PHEV) are able to achieve the highest possible power densities. Each storage energy device has a different model. Several control approaches are applied to control the energy storage devices. In [8, 9], model predictive control (MPC) is presented for residential energy systems with photovoltaic (PV) system and batteries. Model predictive control predicts the load and the generation over a certain time horizon into the future and finds the optimum schedule of the battery over that period which can minimize a desired objective. In [10], a voltage regulation in distribution feeders is proposed using residential energy storage units. The control method is carried out by making the charging and discharging rates of the batteries a function in the voltage at the point of common coupling. A fuzzy logic based control method of battery state of charge (SOC) is presented in [11]. This control method regulates the battery SOC at expected conditions, and consequently the energy capacity of BESS can be small. In [12], a state-of-charge feedback control technique is used to keep the charging level of the battery within its proper range while the battery energy storage system make the output fluctuation of a wind farm smooth. The optimal design of MPC with SMES based on the bat-inspired algorithm (BIA) is introduced for load frequency control in [13]. This work is extended to include the MPC with SMES and CES in [14]. Decentralized MPC with PHEVs is utilized for frequency regulation in a smart three-area interconnected power system in [15].

2. Superconducting magnetic energy storage

The SMES units are used to compensate the load increments by the injection of a real power to the system and diminished the load decrements by the absorbing of the excess real power via large superconducting inductor [16–18]. **Figure 1a** show a schematic diagram of SMES unit consists of superconducting inductor (L), Y-Y/Δ transformer, and controlled ac/dc bridge converter with 12-pulse thyristor. A power conversion system (PCS) is used to connect the superconducting inductor with the AC grid. The PCS is a dual-mode converter and it works as a rectifier or as an inverter in the charging and discharging modes of the inductor respectively. Obviously, the mode of operation is detected according to the nature of load perturbation. The charging phase represents the rectifier mode. In the rectifier mode, adjusted positive voltage is applied across the terminals of the inductor. Alternatively, the discharging phase represents the inverter mode. In the inverter mode, adjusted negative voltage is applied across the terminals of the inductor. The controlling in the thyristor firing angle is used to switch either rectifier or inverter modes, the converter output voltage is expressed in kV and it is given in following equation [19]:

$$E_d = 2E_o \cos(\alpha) - 2I_d R_c \tag{1}$$

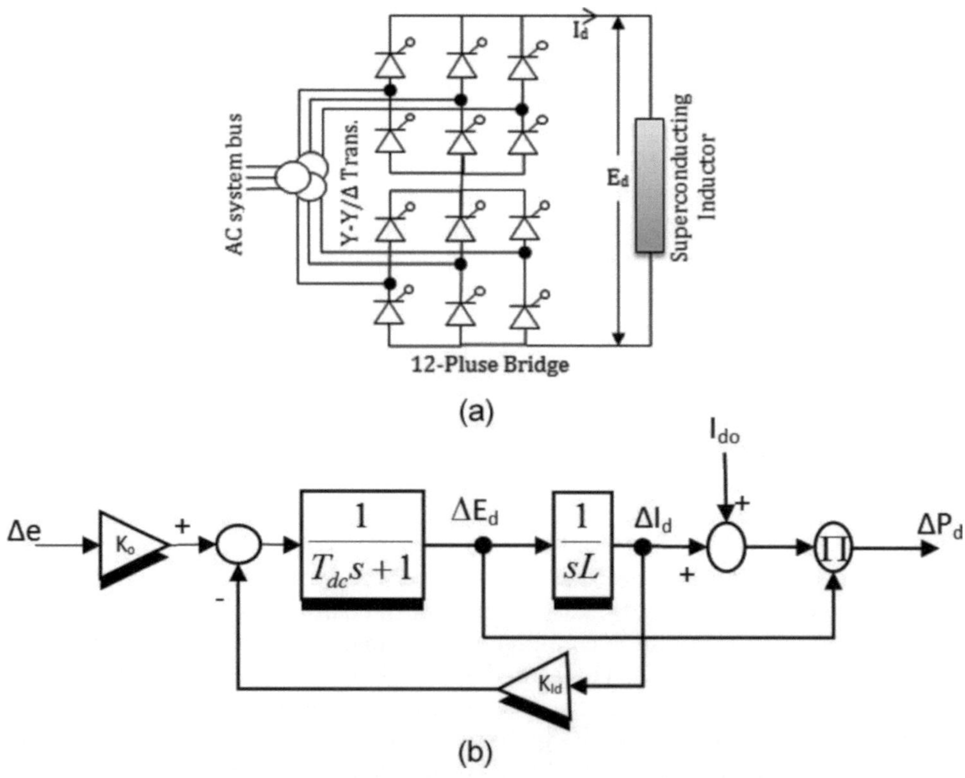

Figure 1.
The SMES unit (a) circuit diagram and (b) corresponding block diagram.

where E_d is the inductor DC voltage (kV); E_o is the converter open circuit voltage (kV); α is the thyristor firing angle (degrees); I_d is the inductor current (kA); R_C is the equivalent resistance of commutation (ohm).

2.1 Modeling of superconducting magnetic energy storage

According to the rectifier or inverter modes, the polarity of the voltage E_d is adjusted while the direction of inductor current I_d does not change. As mention in the above section, the regulation of the thyristor firing angle is used to controlling the direction and magnitude of the inductor power P_d. Initially, a small positive voltage is applied to charge the inductor to its rated current according to the desired charging period of the SMES unit. The inductor voltage is reduced to zero and the inductor current reach to its rated value because the coil is superconducting. When the inductor current reached to its rated value, the SMES unit can be coupled to the power system. The error signal Δe represent the input to control the SMES voltage E_d. This error signal may be the change of system frequency, the change of system voltage, or the change of system current according to the control object. The incremental voltage and current changes of the SMES coil are given as follows:

$$\Delta E_d = \frac{K_o}{1 + sT_{dc}} (\Delta e) \tag{2}$$

$$\Delta I_d = \frac{1}{sL} \Delta E_d \tag{3}$$

where T_{dc} is the converter time constant in sec, K_o is the gain of the proportional controller in kV/Hz and s is the Laplace operator. As reported in [19], the inductor current reaches to its nominal value very slowly in the SMES unit. So, the fast rate of the current to restore its rated value is required to extinguish the next load perturbation fastly. Therefore, a negative feedback signal is used in the SMES control loop to provide fast current recovery. Thus, Eq. 2 is rewritten in following form:

$$\Delta E_d = \frac{1}{1 + sT_{dc}} (K_o \Delta e - K_{Id} \Delta I_d) \tag{4}$$

where $K_{Id}$is the negative feedback gain of the current deviation (kV/kA). In the storage mode, the coil is short-circuited, i.e. $E_{do} = 0$ and there is no power transfer. So in either phase (charging/discharging), the power is defined by $P_d = E_d I_d$ and the initial inductor power is $P_{do} = E_{do} I_{do}$, where E_{do} and I_{do} are the voltage and current magnitudes previous to load disturbance. The inductor power following to the load disturbance is defined as follows:

$$\begin{aligned} P_d &= (E_{do} + \Delta E_d)(I_{do} + \Delta I_d) \\ &= E_{do} I_{do} + E_{do} \Delta I_d + I_{do} \Delta E_d + \Delta E_d \Delta I_d = P_{do} + E_{do} \Delta I_d + I_{do} \Delta E_d + \Delta E_d \Delta I_d, \; E_{do} = 0 \end{aligned} \tag{5}$$

Therefore, the real power incremental change ΔP_d of the SMES unit in MW is computed as follows:

$$\Delta P_d = P_d - P_{do} = I_{do} \Delta E_d + \Delta I_d \Delta E_d \tag{6}$$

The corresponding block diagram of an SMES incorporating the negative feedback of the current deviation is shown in **Figure 1b**.

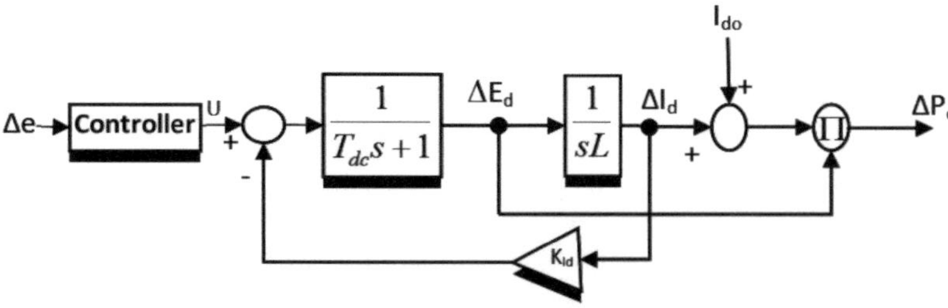

Figure 2.
The block diagram of SMES with controller.

Setting the parameters (L, K_o, K_{Id} and I_{do}) of the SMES unit to their optimistic values can enhance its role in achieving well-damped to the responses. Herein, the application of artificial intelligence (AI) techniques is suggested to search for the optimal parameters of the SMES and controller simultaneously.

2.2 Control techniques of SMES

Modern control techniques such as adaptive control, fuzzy logic control, and model predictive control (MPC) can be applied to control the charging and discharging of the SMES instead of the proportional controller as shown in **Figure 2**. The controller and SMES parameters must be adjusted by proper optimization technique such as genetic algorithm (GA), particle swarm optimization (PSO), and artificial bee colony (ABC),...etc. to give a good performance.

3. Capacitive energy storage

The capacitive energy storage (CES) has an important role to stabilize the power system against to the sudden change in load demand. The static operation of the CES makes its response faster than of the mechanical systems [20–23]. Parallel storage capacitors form the CES. **Figure 3a** shows a schematic diagram of a CES unit connected with the AC grid by a PCS. The capacitor bank dielectric and leakage losses are defined by the resistance (R). When the load demand decreases, the capacitor charges up to its rated full value, thus releases an amount of the excess energy in the system. Contrary, the capacitor discharges to its initial value of voltage when the load demand rises suddenly and release the stored energy fastly to the grid through the PCS. A gate turn-off (GTO) thyristors is used as switches to control the direction of the capacitor current during the charging and discharging as shown in **Table 1**.

The controlling in the thyristor firing angle is used to switch either rectifier or inverter modes of CES to adjust the capacitor voltage as defined in Eq. 1.

3.1 Modeling of superconducting magnetic energy storage

The CES unit is ready to be coupled to the power system for LFC when the rated voltage across the capacitor is attained. The current I_d of CES is controlled by sensing the error signal Δe. This error signal may be the change of system frequency, the change of system voltage, or the change of system current according to the control object. The incremental current changes of the CES are given as follows:

$$\Delta I_d = \frac{K_c}{1 + sT_{dc}}(\Delta e) \tag{7}$$

where K_c is the proportional controller in kA/Hz.

As stated in [20], the CES voltage reaches to its nominal value very slowly. So, the fast rate of the capacitor voltage to restore its rated value is required to extinguish the next load perturbation fastly. Therefore, a negative feedback signal is used in the CES control loop to provide a fast voltage recovery. Thus, Eq. 7 is rewritten in following form:

$$\Delta I_d = \frac{1}{1 + sT_{dc}}(K_c\Delta f - K_{Ed}\Delta E_d) \tag{8}$$

where K_{Ed} is the negative feedback gain of the capacitor voltage deviation (kA/kV). In the storage mode, the capacitor represents an open circuit, i.e. $I_{do} = 0$

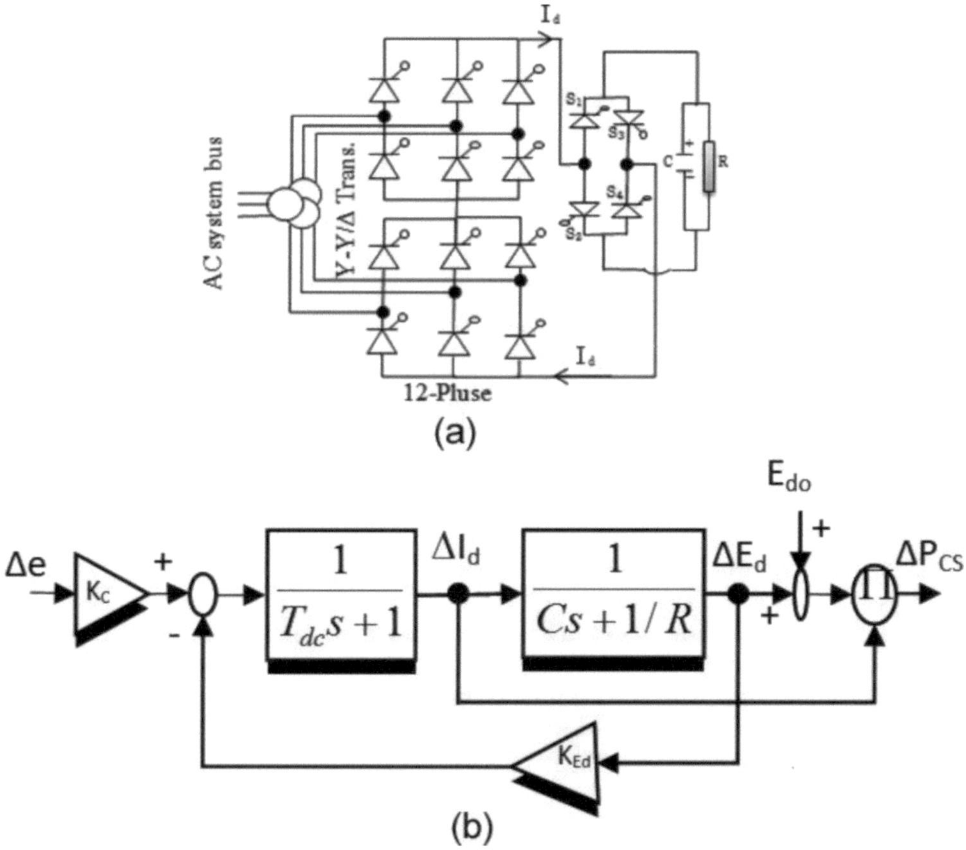

Figure 3.
The CES unit (a) circuit diagram and (b) corresponding block diagram.

	Charging mode	Discharging mode
S_1, S_4	ON	OFF
S_2, S_3	OFF	ON

Table 1.
The modes of switches during the charging and discharging of CES unit.

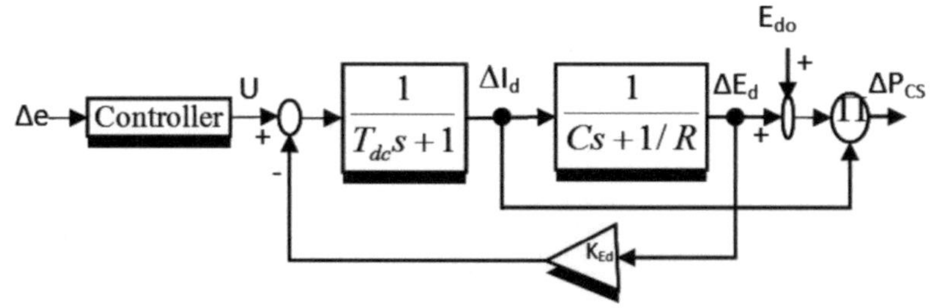

Figure 4.
The block diagram of CES with controller.

and no power transfer. Hence, the power is defined by $P_{CS} = E_d I_d$ and the initial CES power is $P_{CSo} = E_{do} I_{do}$, where E_{do} and I_{do} are the magnitudes of the voltage and current prior to load disturbance. Following a load disturbance, the power flow into the CES is given as follows:

$$\begin{aligned} P_{CS} &= (E_{do} + \Delta E_d)(I_{do} + \Delta I_d) \\ &= E_{do} I_{do} + E_{do}\Delta I_d + I_{do}\Delta E_d + \Delta E_d \Delta I_d = P_{CSo} + E_{do}\Delta I_d + I_{do}\Delta E_d + \Delta E_d \Delta I_d,\ I_{do} = 0 \end{aligned} \tag{9}$$

Thus, the real power incremental change ΔP_{CS} of the CES unit in MW is computed as follows:

$$\Delta P_{CS} = P_{CS} - P_{CSo} = E_{do}\Delta I_d + \Delta E_d \Delta I_d \tag{10}$$

The corresponding block diagram of a CES unit incorporating the negative feedback of the voltage deviation is shown in **Figure 3b**.

Setting the parameters (C, K_c, K_{Ed} and E_{do}) of the CES unit to their optimistic values can enhance its role in achieving well-damped to the responses. Herein, the application of artificial intelligence (AI) techniques is suggested to search for the optimal parameters of the CES and controller simultaneously.

3.2 Control techniques of CES

Modern control techniques such as adaptive control, fuzzy logic control, and MPC can be applied to control the charging and discharging of the CES instead of the proportional controller as shown in **Figure 4**. The controller and CES parameters must be adjusted by proper optimization technique such as GA, PSO, and ABC, ...etc. to give a good performance.

4. Plug-in hybrid electric vehicle model

The PHEV model is represented as first-order transfers function with very small time constant T_{PHEV} as shown in **Figure 5(a)** [24–27]. The change of PHEV output power ΔP_{PHEV} for charging or discharging is selected according to the control signal U_i of the controller. In this chapter, the control signal is determined by modern control techniques such as adaptive control, fuzzy logic control, and model predictive control (MPC). The control signal depends on the error signal to adjust the charging or discharging of PHEVs batteries. **Figure 5(b)** shows a bi-directional PHEV to charging and discharging power control 'vehicle to grid (V2G)'. According to the change of error, this V2G release a power to the grid or drain power from the

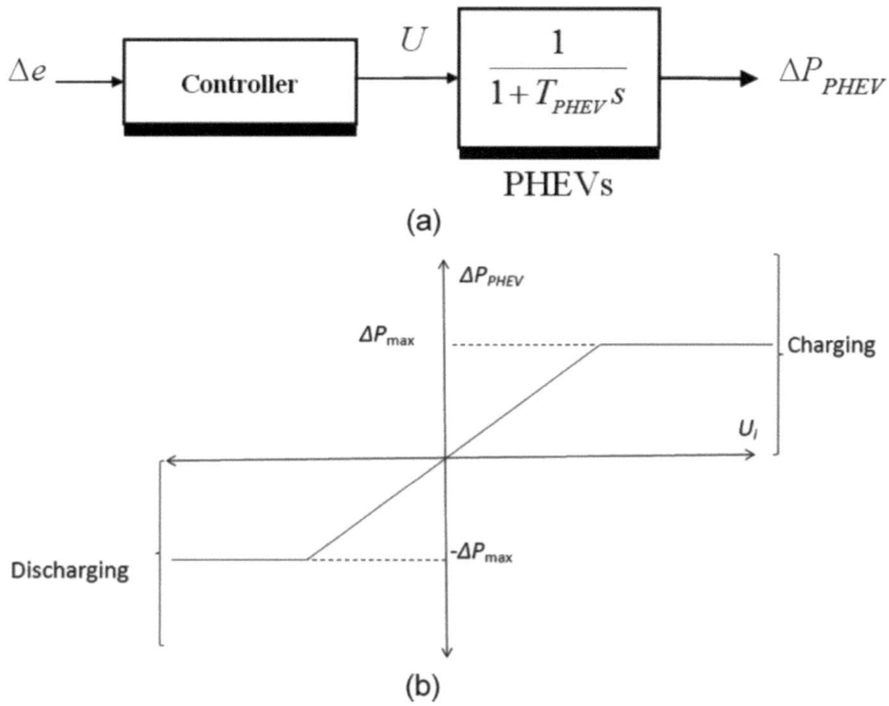

Figure 5.
PHEV model: (a) PHEV with controller block diagram, (b) change of PHEV output power against control signal.

grid. The change of PHEV output power is adjusted by the control signal (U) according to the limit range of output power deviation of PHEV as

$$\Delta P_{PHEV} = \begin{cases} U_i, \ |U_i| \le \Delta P_{\max} \\ \Delta P_{\max}, \ U_i > \Delta P_{\max} \\ -\Delta P_{\max}, \ U_i < -\Delta P_{\max} \end{cases} \tag{11}$$

where $P_{\max}$ is the maximum PHEV power.

5. Conclusion

In this chapter, classifications of energy storage devices and control strategy for storage devices by adjusting the performance of different devices and features of the power imbalance are presented. The modeling of each storage energy devices is discussed. Furthermore, the control method for each one are cleared. These energy storage devices with modern control techniques such as adaptive control, fuzzy logic control, and model predictive control (MPC) can be applied to extinguish the rapid change in load demands and the fluctuations of RES.

Author details

Mahmoud Elsisi
Electrical Power and Machines Department, Faculty of Engineering (Shoubra), Benha University, Cairo, Egypt

*Address all correspondence to: mahmoud.elsesy@feng.bu.edu.eg

References

[1] Heide D, Greiner M, Von Bremen L, Hoffmann C. Reduced storage and balancing needs in a fully renewable European power system with excess wind and solar power generation. Renewable Energy. 2011;**36**(9): 2515-2523

[2] Yamegueu D, Azoumah Y, Py X, Kottin H. Experimental analysis of a solar PV/diesel hybrid system without storage: Focus on its dynamic behavior. International Journal of Electrical Power & Energy Systems. 2013;**44**(1):267-274

[3] Datta M, Senjyu T, Yona A, Funabashi T. Frequency control of photovoltaic–diesel hybrid system connecting to isolated power utility by using load estimator and energy storage system. IEEJ Transactions on Electrical and Electronic Engineering. 2010;**5**(6): 677-687

[4] Dıaz-Gonzalez F, Sumper A, Gomis-Bellmunt O, Villafafila-Robles R. A review of energy storage technologies for wind power applications. Renewable and Sustainable Energy Reviews. 2012; **16**(4):2154-2171

[5] Bandara K, Sweet T, Ekanayake J. Photovoltaic applications for offgrid electrification using novel multi-level inverter technology with energy storage. Renewable Energy. 2012;**37**(1):82-88

[6] Zhou Z, Benbouzid M, Fredéric Charpentier J, Scuiller F, Tang T. A review of energy storage technologies for marine current energy systems. Renewable and Sustainable Energy Reviews. 2013;**18**:390-400

[7] Di Fazio A, Erseghe T, Ghiani E, Murroni M, Siano P, Silvestro F. Integration of renewable energy sources, energy storage systems, and electrical vehicles with smart power distribution networks. Journal of Ambient Intelligence and Humanized Computing. 2013;**4**:663-671

[8] Schreiber M, Hochloff P. Capacity-dependent tariffs and residential energy management for photovoltaic storage systems. In: Power and Energy Society General Meeting (PES), IEEE. 2013. pp. 1-5

[9] Worthmann K, Kellett CM, Braun P, Grüne L, Weller SR. Distributed and decentralized control of residential energy systems incorporating battery storage. IEEE Transactions on Smart Grid. 2015;**6**(4):1914-1923

[10] Wang Y, Wang BF, Son PL. A voltage regulation method using distributed energy storage systems in lv distribution networks. In: Energy Conference (ENERGYCON), IEEE. 2016. pp. 1-6

[11] Li X, Hui D, Wu L, Lai X. Control strategy of battery state of charge for wind/battery hybrid power system. In: IEEE International Symposium on Industrial Electronics (ISIE). 2010. pp. 2723-2726

[12] Yoshimoto K, Nanahara T, Koshimizu G, Uchida Y. New control method for regulating state-of-charge of a battery in hybrid wind power/battery energy storage system. In: Power Systems Conference and Exposition, IEEE PES. 2006. pp. 1244-1251

[13] Elsisi M, Soliman M, Aboelela MAS, Mansour W. Optimal design of model predictive control with superconducting magnetic energy storage for load frequency control of nonlinear hydrothermal power system using bat inspired algorithm. Journal of Energy Storage. 2017;**12**:311-318

[14] Elsisi M, Soliman M, Aboelela MAS, Mansour W. Improving the grid frequency by optimal design of model

predictive control with energy storage devices. Optimal Control Applications and Methods. 2018;**39**(1):263-280

[15] Elsisi M, Soliman M, Aboelela MAS, Mansour W. Model predictive control of plug-in hybrid electric vehicles for frequency regulation in a smart grid. IET Generation, Transmission & Distribution. 2017;**11**(16):3974-3983

[16] Ali MH, Wu B, Dougal RA. An overview of SMES applications in power and energy systems. IEEE Transactions on Sustainable Energy. 2010;**1**(1):38-47

[17] Bhatt P, Roy R, Ghoshal SP. Comparative performance evaluation of SMES–SMES, TCPS–SMES and SSSC–SMES controllers in automatic generation control for a two-area hydro–hydro system. International Journal of Electrical Power & Energy Systems. 2011;**33**(10):1585-1597

[18] Banerjee S, Chatterjee JK, Tripathy SC. Application of magnetic energy storage unit as load-frequency stabilizer. IEEE Transactions on Energy Conversion. 1990;**5**(1):46-51

[19] Tripathy SC, Balasubramanian R, Chandramohanan Nair PS. Adaptive automatic generation control with superconducting magnetic energy storage in power systems. IEEE Transactions on Energy Conversion. 1992;**7**(3):434-441

[20] Tripathy SC. Improved load-frequency control with capacitive energy storage. Energy Conversion and Management. 1997;**38**(6):551-562

[21] Abraham RJ, Das D, Patra A. Automatic generation control of an interconnected power system with capacitive energy storage. International Journal of Electrical Power and Energy Systems Engineering. 2010;**3**(1):351-356

[22] Sanki P, Ray S, Shukla RR, Das D. Effect of different controllers and capacitive energy storage on two area interconnected power system model using Matlab Simulink. In: 2014 First International Conference on Automation, Control, Energy and Systems (ACES). pp. 1-6

[23] Dhundhara S, Verma YP. Evaluation of CES and DFIG unit in AGC of realistic multisource deregulated power system. International Transactions on Electrical Energy Systems. 2016:1-14

[24] Vachirasricirikul S, Ngamroo I. Robust LFC in a smart grid with wind power penetration by coordinated V2G control and frequency controller. IEEE Transactions on Smart Grid. 2014;**5**(1): 371-380

[25] Pahasa J, Ngamroo I. PHEVs bidirectional charging/discharging and SoC control for microgrid frequency stabilization using multiple MPC. IEEE Transactions on Smart Grid. 2015;**6**(2): 526-533

[26] Pahasa J, Ngamroo I. Coordinated control of wind turbine blade pitch angle and PHEVs using MPCs for load frequency control of microgrid. IEEE Systems Journal. 2016;**10**(1):97-105

[27] Pahasa J, Ngamroo I. Simultaneous control of frequency fluctuation and battery SOC in a smart grid using LFC and EV controllers based on optimal MIMO-MPC. Journal of Electrical Engineering & Technology. 2017;**12**(2): 601-611

Chapter 4

Modeling of a Contact-Less Electric-Vehicle Battery-Charging Station Fed from On-Grid Photovoltaic Arrays

Essamudin Ali Ebrahim

Abstract

Electric vehicles (EVs) are environmental friendly due to no exhaust gases or carbon dioxide. In addition, there is no noise through operation. However, up to now, there are some challenges that facing its spread through all over the world. The main problem that these vehicles face is the fast charging process of the used batteries through neat and clean source without plugs. So, this chapter deals with a proposed method for a contactless battery charger of both electric and hybrid electric vehicles (HEVs) from renewable resources. The chapter proposes a public station for fast charging. This station implies off-board battery charger fed from on-grid (OG) photovoltaic (PV) arrays through inductively power transfer (IPT). This comfortable 100 kW contactless power station is designed, modeled, and simulated as a general software package reliable to be used for any other station design. The air gap of the air-core transformer (ACT) divides the station into two parts. The first part implies roof-mounted PV array with its intelligent-controlled maximum power point tracking (MPPT) technique, power converter, three-level power inverter, and resonant compensator converter that operates at high frequency. The second one includes rectifiers and switched mode power converter with smart controller. The chapter includes samples for simulation results that are obtained from the Matlab software package.

Keywords: contactless power transfer, inductive power transfer (IPT), air-core power transformer (ACPT), electric vehicle (EV), photovoltaic (PV), three-level inverter

1. Introduction

Several research works have been introduced to contribute toward introducing an environmentally friendly power source to replace the conventional internal combustion engine (ICE). Electric vehicles (EVs) and fuel cell electric vehicles (FCEVs) expected to be replaced soon. It is known that the EVs have higher efficiency for the energy conversion, capability for regenerative braking and motoring, minimum emission for exhaust gases, and low noise and vibration when compared with ICEs.

The host battery, its total cost, and the development of its performance play an important role in the electric vehicle (EV) and hybrid EV [1]. The battery charging process needs several considerations such as fastness, safety, easy, and low-priced cost.

There are two techniques for charging process: plug-in and un-plugged but connected. Out-door plug-in needs to have a large cable that creates power losses, inconvenience, and electric-chock hazards for most owners. But, on the other hand, inductively coupled power transfer (ICPT) contactless-charging system introduces an imperious solution for the risk and embarrassment problems through transferring the charging energy over a softly inductive coupler [2–4].

Through the last decade, several aspects of ICPT have been studied, such as magnetic coupler design techniques, compensation topologies, control methods, foreign object detection algorithms, and the radiation safety issues [5–7]. Also, the main power supply for the station prefers to be renewable and friend to the environment. High-rating good quality and low-priced photovoltaic (PV) cells – as a renewable energy supply – are nowadays available and suitable to this purpose [8]. Several efforts for progressing the ICPT charging technique are done as described by Miskovski and Williamson [9]. The authors suggested that a 100-kW public solar power station – with battery banks – for partial charging (30%) only and 10% of the required power is supplied from the PV array. The main disadvantages of that system are that it uses a battery bank with bulky and expensive elements for the public-charging station. Also, in Ref. [10], the author presented a charge controller of solar photovoltaic panel (SPV) fed battery. But in [11], Robalino et al. suggested a solar energized docking charging with fuel cell for electric vehicle. The study proposed an alternative method for manufactures and consumers to use the clean sources for transportation commitments. This study has limited theoretical guides for the designers to model the station.

However, this chapter introduces a study that came to provide an integrated model for a solar-powered electric vehicle charging station connected to the main grid. This chapter introduces a complete modeling and designing for a public station of electric vehicle battery charging with a wireless technique. This technique depends on inductively coupled power transfer (ICPT) by using air-core transformer (ACT) to overcome the problems of contacts and wiring due to high-rating charging current. The proposed station is fed from a renewable power resource such as PV arrays connected with the main grid. The study includes the design for the PV array with its maximum power point tracking (MPPT) and its mathematical model. Also, this chapter introduces a Matlab/m-file software package that includes a model for boost converter required for interconnecting between PV array and load with its smart control. In addition, the package introduces a complete design and model for three-level inverter interconnecting between the PV and the main grid. This chapter also proposes an air-core transformer with light weight and a few number of turns for inductively coupling. The proposed system utilizes a buck converter for battery charger to control the charging current with an intelligent PI-tuning controller. In addition, a high-frequency resonant converter with series compensation is modeled and simulated through this study. The validity of the system is verified by charging four 19-kW lithium-ion batteries of Honda EV in order to test the capacity and efficiency of the station.

2. The proposed ICPT-PV for EV public-charging station

The proposed public-charging station consists of two main parts (as shown in **Figure 1**): the static and dynamic sections. The rated power of the station is 100 kW, and it depends on the types and capacity of the EV batteries to be charged.

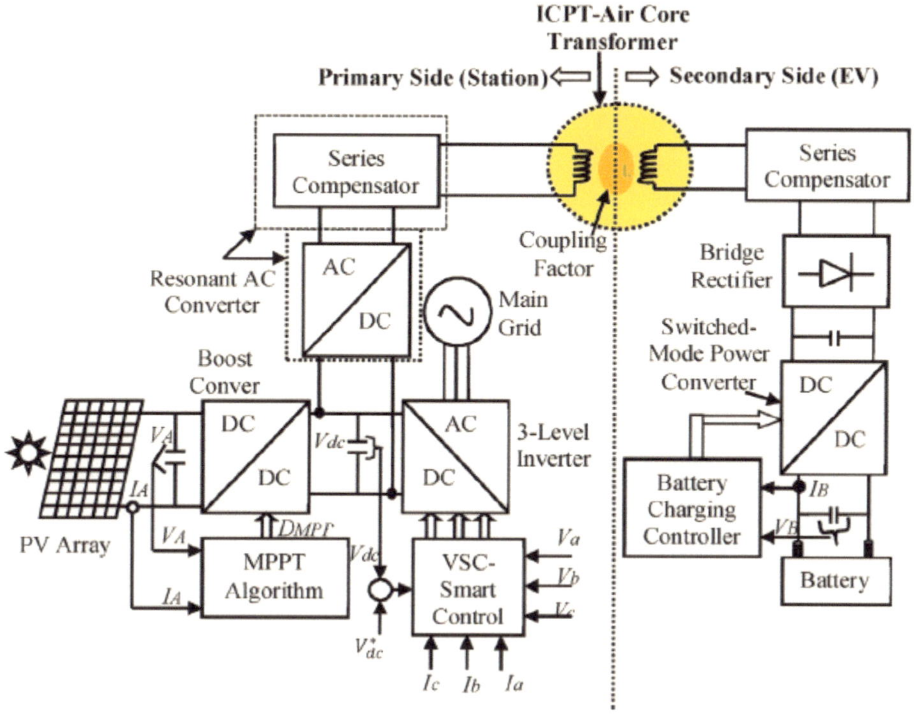

Figure 1.
The proposed public station for PV battery charging.

The average capacity for most batteries used in electric vehicles is around 20 kWh. The parts of the station are classified according to the position that is placed. Static part is the part that placed on land or under ground in the station. But, the dynamic part is that placed inside the EV as on-board battery charging. Each part of the station will be explained and designed in detail in the following sections.

3. The primary-side part

The primary-side part consists of the modeled PV panels in arrays as a neat-renewable energy source with its controller for the maximum power point tracking (MPPT) that implies one boost converter, a bi-directional three-level inverter for interfacing with the main grid and the PV arrays, and resonant converter with very high frequency including a series compensator, and all are connected with the air-core transformer (ACT) through its primary winding.

3.1 Modeling of the PV array with MPPT algorithm

The PV cell, module, and arrays mathematical models can be obtained by Essamudin [8] according to the following equations:

The total current of the cell is computed as follows:

$$I = I_{irr} - I_o\left[e^{\left(\frac{q(V+IR_s)}{nkT}\right)} - 1\right] - \frac{V + IR_s}{R_p} \qquad (1)$$

Also, the total current of the module can be computed from the following equation:

$$I_M = I_{irr} - I_o\left[e^{\left(\frac{q(V_M + I_M N_s R_s)}{N_s nkT}\right)} - 1\right] - \frac{V_M + I_M N_s R_s}{N_s R_p} \tag{2}$$

Then, the total PV array current is given by the following criteria:

$$I_A = N_p I_{irr} - I_o\left[e^{\left(\frac{q\left(V_A + I_A \frac{N}{N_p} R_s\right)}{N_s nkT}\right)} - 1\right] - \frac{V_A + I_A\left(\frac{N}{N_p}\right)R_s}{\frac{N_s}{N_p}R_p} \tag{3}$$

where I, I_M, I_A are the cell, module, and array currents (A); I_{irr} is the irradiance or photo current (A); I_o is the diode saturation current (A); V, V_M, V_A are the cell, module, and array voltages (V); $q = 1.6 \times 10^{-19}$ (C); $K = Boltizmann\ constant = 1.3806503 \times 10^{-23}$J/K; R_s, R_p are series and parallel resistance (Ω); N_s, N_p are series and parallel cell numbers; and T is the cell temperature (°K).

The proposed station is designed to work as on-grid with 100-kW power rating supplied from the PV source in the day light (not using battery banks as storage elements), and it will be supplied from the main grid in the evening and cloud days. So, the following lines show the modeling and designing for the number of PV modules with arrays to generate the power needed.

The rated power for the station is 100 Kw, and the rated output voltage for the PV array is 300 V, if it is supposed that the SunPower module (SPR-305) types [12–14] are used with their characteristics: voltage for open circuit (V_{oc}) is 64.2 V, current for short circuit (I_{sc}) is 5.96 A, maximum power voltage and current V_{MPP} is 54.7 V and I_{MPP} is 5.58 A, $\therefore N_s = \left(\frac{300}{64.2}\right) = 4.6 \cong 5$ modules, then the DC output terminal voltage of the PV array will be 320 V.

If the maximum power for one module equals (54.7×5.58)≅ 305 W, then the total required number of modules can be obtained as $\left(\frac{100000}{305}\right) = 327$ modules = $N_s \times N_p$; dependently, if the total numbers of the parallel modules are calculated as (327/5) ≅ 66 modules, then the total required modules for the proposed system are ($66 \times 5 = 330$), and the maximum power will be equal to P_{MPP}=(330×305)/1000 = 100.65 kW.

When the dimension of one proposed module is 1.559m × 1.046m = 1.63m^2 [14], then, the total surface area required is equal (1.63×330≅538m^2). This area can be saved on the station roof and used to arrange the modules in rows and columns. However, the characteristic curves of one module and one array for the V-I and V-P curves are illustrated in **Figure 2** (**a** and **b**), respectively.

3.2 Maximum power point tracking (MPPT) algorithm with boost converter

There are several methods proposed to achieve MPPT [15, 16]. The simplest method is the incremental conductance (IC) that depends on the base that the maximum power point is obtained when the power slope of the PV is zero that obtained from the derivative of the power due to the voltage and equating it to zero (i.e., $dP/dV = 0$). **Figure 2a** and **b** describes this and explores the characteristic curves of both power and current with the PV terminal voltage of one module and one array. As shown, the left side of the curve is positive and negative in the right. Due to this condition, the MPP can be found in terms of the increment in the array conductance. In Eq. (4), if the value of the error in the right hand side equal zero, this means that the change of current due to voltage equal to negative value of the

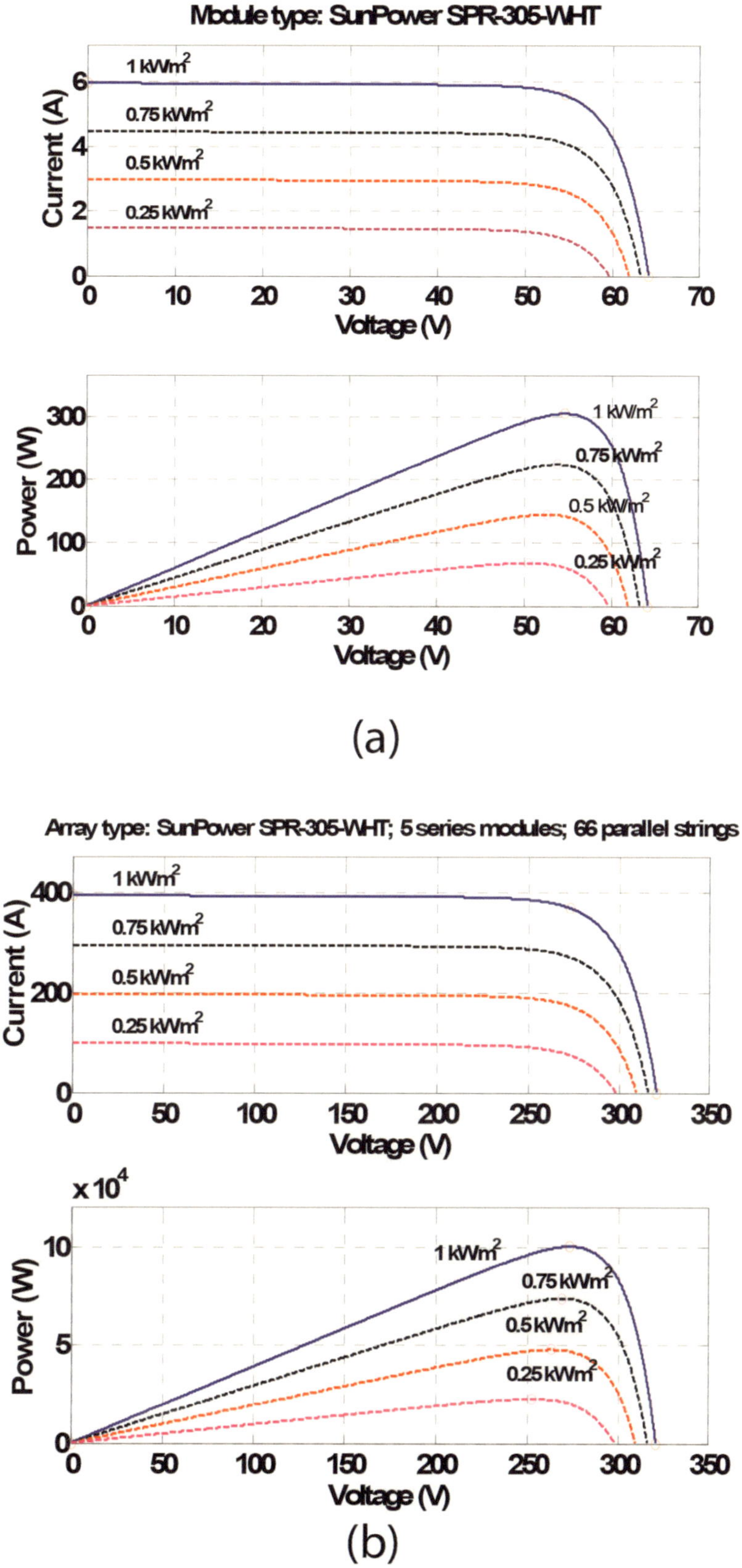

Figure 2.
V-I and V-P characteristics for (a) one module and (b) the PV array.

division of current by voltage. The quotient of this division equal to the incremental conductance (IC).

$$\frac{dP_A}{dV_A} = \frac{d(V_A.I_A)}{dV_A} = I_A + V_A\frac{dI_A}{dV_A} = \text{error} \tag{4}$$

The IC has three states that can be presented by the following Eqs. From Eq. (5)–(7) [15].

$$\frac{\Delta I_A}{\Delta V_A} = -\frac{I_A}{V_A} = -G \tag{5}$$

or

$$\frac{\Delta I_A}{\Delta V_A} > -\frac{I_A}{V_A} \tag{6}$$

or

$$\frac{\Delta I_A}{\Delta V_A} < -\frac{I_A}{V_A} \tag{7}$$

At MPP, the right-hand side (RHS) of the equation equals to the left-hand side (LHS; Eq. 5). If the LHS is greater than RHS (Eq. 6), the power point lies on the left. If Eq. (7) is verified, the point lies in the right of MPP. This method is similar to the Perturb and Observe method for searching the MPPT. This method has good power transient performances through the atmospheric conditions rapid changes.

The error signal (e) can be minimized by using an integral regulator, then the duty ratio at MPP (D_{MPP}) is obtained as:

$$D_{MPP} = K_I\int_0^{\Delta t}\left(\frac{\Delta I_A}{\Delta V_A} + \frac{I_A}{V_A}\right)dt \tag{8}$$

The final output control signal needed to trigger the IGBT boost converter switch can be obtained with the help of Matlab/Simulink as illustrated in **Figure 3**.

3.3 Boost converter

The main function of boost converter is used to boost voltage, so it is used with PV to raise its output voltage from 320 to 540 V. The main components are a series inductor with shunt capacitor, and both are controlled through the electronic

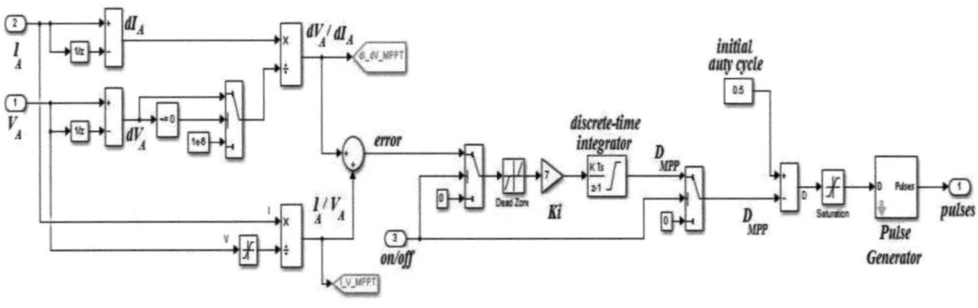

Figure 3.
Matlab/Simulink MPPT algorithm with boost-converter pulse generation.

switches such as MOSFET or IGBT (as shown in **Figure 4**). The main triggering signals are generated as train of pulses from MPPT-algorithm output when the duty-ratio varies.

3.4 Soft-switching multi-level three-phase diode-clamped inverter topology

The three-phase multi-level (with three-level) diode-clamped inverter that proposed in this research is shown in **Figure 5**. The main idea depends on the topology of the neutral point clamped (NPC), which was proposed by Nabae, Takahashi, and Akagi in 1981 [17, 18].

To interconnect between the DC output of the PV and the main AC-grid, it should be used as a bi-directional converter to transfer power from DC-side to AC-one and vice versa. So, a three-level voltage-source current-controlled converter with neutral point clamped (NPC) is proposed. **Figure 5** shows the circuit configuration of the NPC inverter. Each leg has four IGBTs connected in series. The applied voltage on the IGBT is one-half of the conventional two-level inverter. The bus voltage is split into two by the connection of equal series connected bus capacitors. Each leg is completed with the addition of two clamp diodes. The main advantages of this technique are its capability of handling higher voltages, lower line-to-line and common-mode voltage steps, and lower output current ripple for the same switching frequency as that used in a two-level inverter.

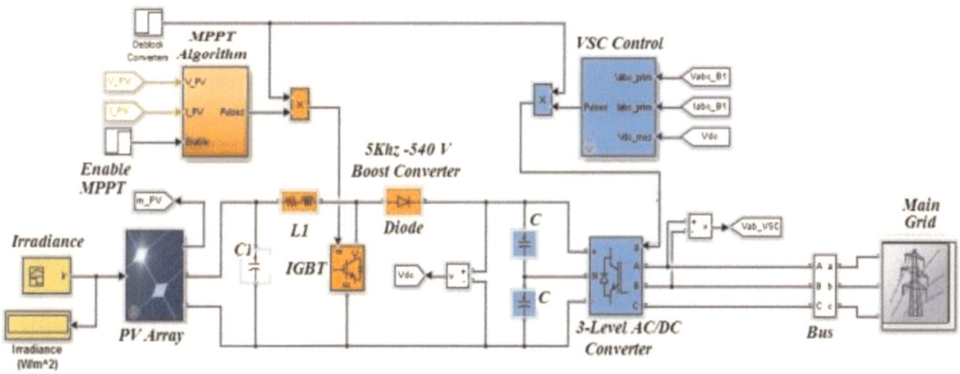

Figure 4.
Modeling of on-grid PV array with MPPT converter and inverter set.

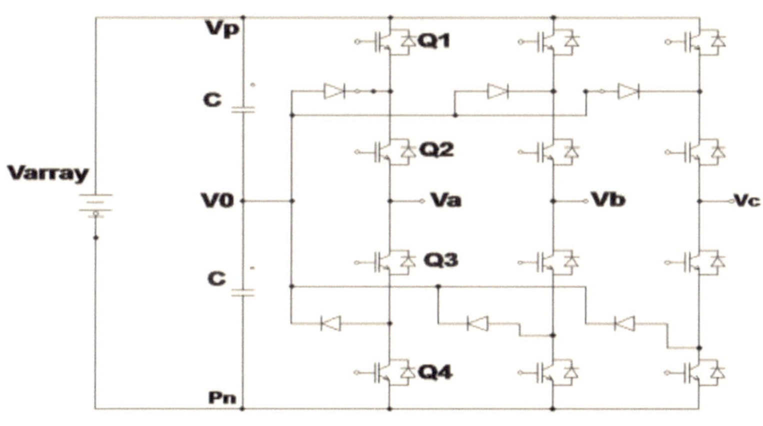

Figure 5.
Three-phase three-level inverter (power circuit).

This converter can produce three voltage levels on the output: the DC bus plus voltage (V_p), zero voltage (V_0), and DC bus negative voltage. For a one phase (phase A) operation, when IGBTs Q_1 and Q_2 are turned on, the output is connected to V_p; when Q_2 and Q_3 are turned on, the output is connected to V_0; and when Q_3 and Q_4 are turned on, the output is connected to V_n [19].

3.5 Voltage-source DC/AC converter/inverter control

The boost-converter output should be constant and equal or greater than the peak value of the AC supply voltage for the grid. To do this, a robust controller should be used. The proposed technique uses PI-smart controller with self-tuning adaptive (K_p, K_i) parameters by using bacteria-foraging optimization algorithm (BFO). The voltage error signal e_v is the difference between the actual and reference DC-link voltage $\left(e_v = V_{dc}^* - V_{dc}\right)$. This value should be minimized to a small or zero [8]. The DC-link voltage can be computed as the value that equals to peak value of the line-to-line voltage ($380\sqrt{2} \cong 540$ V). In the same time, the inverter power factor (P.F.) is equal to unity. This can be achieved by using another PI-smart controller for the current and making $I_{qs}^* = 0$. This controller is used to minimize the current error signal that is the difference between the direct reference and the actual current components $\left(e_i = I_{ds}^* - I_{ds}\right)$. **Figure 6** illustrates the overall control algorithm and the Matlab/Simulink blocks. The switching frequency for the converter is equal to 30 KHz to give 50 Hz voltage output.

3.6 Modeling of contactless power-transfer air-core transformer (ACT)

One of the more reliable and safe ways to transfer power is un-pugged but connected wireless method. This method depends on the inductively coupling air-core transformer with a large air gap. The power is energized from the floor of the station to the primary winding, and then by the effect of induction, it is transferred to the secondary winding that puts under the vehicle through a wide air gap. So, this transformer needs special construction to design the primary and secondary windings and its air-gap. There are several configurations for this transformer such as the core of coil: some have air core, and others have iron core. Also, several factors should be taken into consideration through modeling and designing of ACT such as

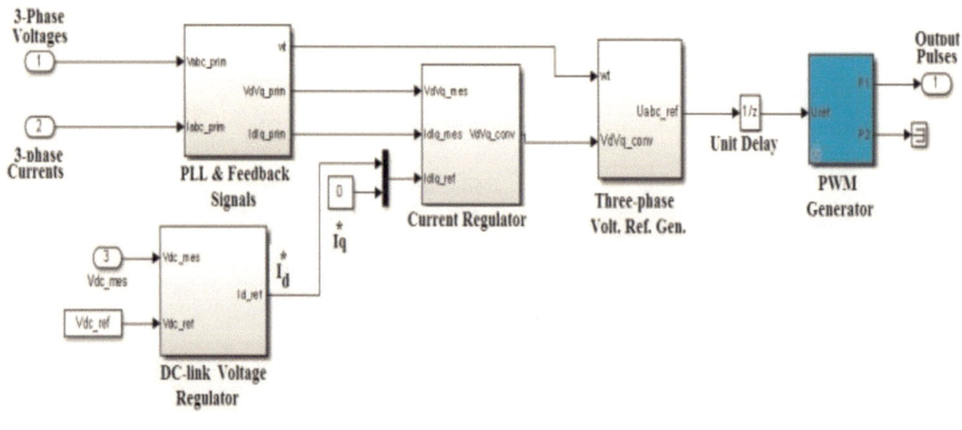

Figure 6.
Control algorithms for DC-link voltage and current regulator for three-level inverter.

the air gap between the primary and secondary winding, cost of the magnetic material for the core, weight of the core, eddy current losses in the core, operating frequency, and sensitivity to misalignment between primary and secondary windings. Due to the large air gap for the transformer, this configuration leads to a large leakage inductance and low mutual coupling that involves a large magnetizing current. Because of CPT transformer with air core has a lightweight with core losses, it is selected in this study for both windings. The coupling coefficient, k, affects directly on the capability of the power transferred of the ICPT. This coefficient can be computed as $k = \frac{M}{\sqrt{L_1 L_2}}$, where L_1 and L_2 are the self-inductance coefficients of the primary and secondary coils, and M is the mutual inductance between them. **Figure 7a** shows the schematic diagram of the transformer, and **Figure 7b** illustrates the equivalent circuit of the model, where L_p and L_s are the leakage inductance of both primary and secondary windings of the ACT. The power-transfer capability to the batteries can be improved by compensation to eliminate the effect of induction. This can be achieved by connecting a capacitor in both sides of the transformer. It is called series-series (SS) compensation as shown in **Figure 7c**. This combination makes the primary capacitance independent of both

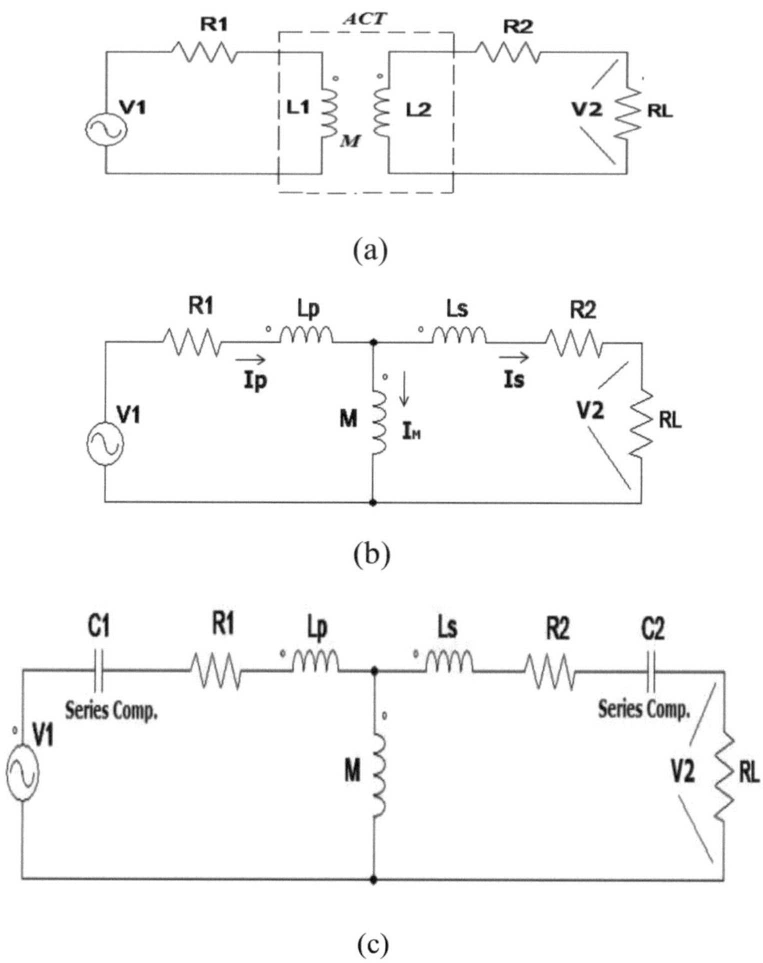

Figure 7.
Modeling of an ICPT air-core transformer with SS compensation. (a) Schematic diagram. (b) Equivalent circuit. (c) Equivalent circuit with SS-compensation.

the magnetic coupling and the load [2]. According to the resonant frequency through both primary and secondary coils, it can compute and select the values of the primary and secondary capacitances C_1 and C_2, respectively. This maximizes the power transferred to the batteries and improves the power factor of the supply to unity.

The power transferred from the primary to the secondary is given as follows:

$$P_2 = \frac{\omega_0^2}{R_L} M^2 I_p^2 \tag{9}$$

where ω_0, I_p, and R_L are the resonant frequency, primary current, and the load resistance, respectively. The value of ω_0 is chosen equal to both primary and secondary windings and can be computed as follows:

$$\omega_0 = \frac{1}{\sqrt{L_p C_1}} = \frac{1}{\sqrt{L_s C_2}} \tag{10}$$

So, the capacitance values of both primary and secondary windings $C_1$and C_2 can be obtained as follows:

$$C_1 = \frac{1}{L_p \omega_0^2} \text{ and } C_2 = \frac{1}{L_s \omega_0^2} \tag{11}$$

Designing of ACT depends on the values of its parameters L_1, L_2, and M. This can be obtained from its physical dimensions. The coils with rectangular form are proposed to both primary and secondary windings of ACT (as shown in **Figure 8**). The advantage of this form is giving much better tolerance to misalignment. The flow chart illustrated in [4] illustrates the design process for the overall ACT, and the parameters of 100 kW charging station are tabulated in **Table 1**.

A general software package with help of the Matlab m-file is introduced for designing and simulation of the ACT. This package is validated with much accuracy for any ACT with a rectangular form for different dimensions and relative position between them.

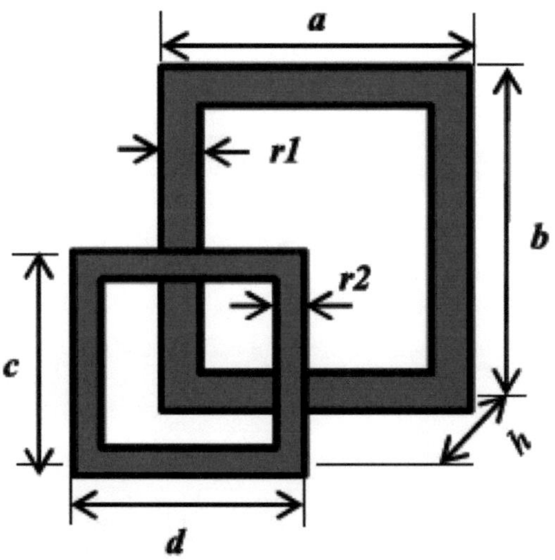

Figure 8.
Geometrical dimensions of the proposed ACT.

Parameter	Value
No. of primary turns (N_1)	6
No. of secondary turns (N_2)	4
Radius of the primary coil (r_1)	9.27 mm
Radius of the secondary coil (r_2)	8.00 mm
The primary-coil length and width *(a,b)*	0.9 m
The secondary-coil length and width *(c,d)*	0.6 m
Internal resistance of the primary (R_1)	1.2365 mΩ
Internal resistance of the secondary (R_2)	2.2781 mΩ
Primary self-inductance (L_1)	0.082 mH
Secondary self-inductance (L_2)	0.041 mH
Mutual inductance between coils (M)	.011 mH
Air gap length	175 mm
The coupling coefficient, k	0.2
Voltage of secondary (V_2)	500 V
Resonant frequency (f_0)	20 KHz
Power of secondary (P_2)	100 kW
Primary series compensation (C_1)	0.7722 uf
Secondary series compensation (C_2)	1.5441 uf

Table 1.
Parameters of ACT.

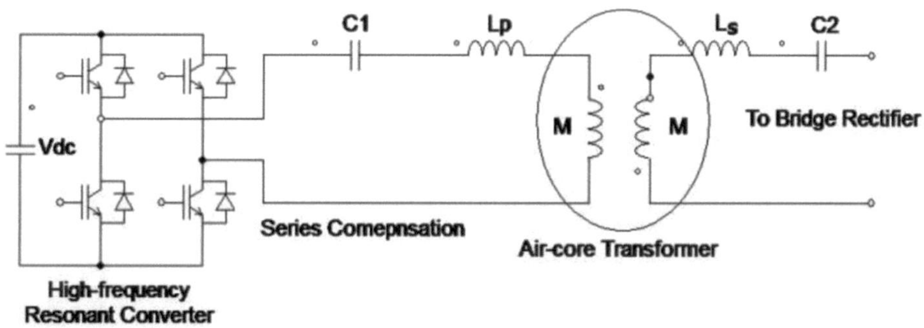

Figure 9.
The high-frequency resonant converter with SS compensation of ACT.

3.7 Resonant DC/AC high-frequency converter

DC-link output voltage cannot be transferred directly to the AC transformer. So, a resonant AC converter is required. This converter is a single-phase bridge inverter switched at a very high-resonant frequency as shown in **Figure 9**. The output of the converter is fed to the primary winding of the air-core transformer through a series compensated capacitor.

4. Second part (mobile side)

The second part of the proposed system implies the secondary coil of the ACT that constructed inside the EV with inductive coupling with primary winding through the air gap. This side (as shown in **Figures 1** and **10**) includes single-phase uncontrolled bridge rectifier, switched-mode power buck converter with its intelligent battery-charging controller, and tracking batteries.

4.1 DC/DC Buck converter

An un-controlled bridge rectifier is required to convert the energized AC secondary-coil power to a DC current. The output voltage of the bridge is greater than the battery-terminal voltage. So, a buck converter (as shown in **Figure 10**) is needed. The main components of it are a controlled switch, diode, series reactor, and shunt capacitor. The proposed smart control circuit is proposed to adapt the battery-bank charging voltage (**Figure 11**).

4.2 Mathematical model of EV battery

The actual mathematical dynamic model of EV battery for simulation will be obtained from Refs. [20, 21] in the following sections:

4.2.1 For lead-acid battery

Charging:

$$V_{bat} = E_0 - R.i - K\frac{Qi^*}{(it - 0.1\,Q)} - K\frac{Qit}{(\,Q - it)} + e^t \tag{12}$$

$$\text{Discharge}: V_{bat} = E_0 - R.i - K\frac{Q(it + i^*)}{(\,Q - i.t)} + e^t \tag{13}$$

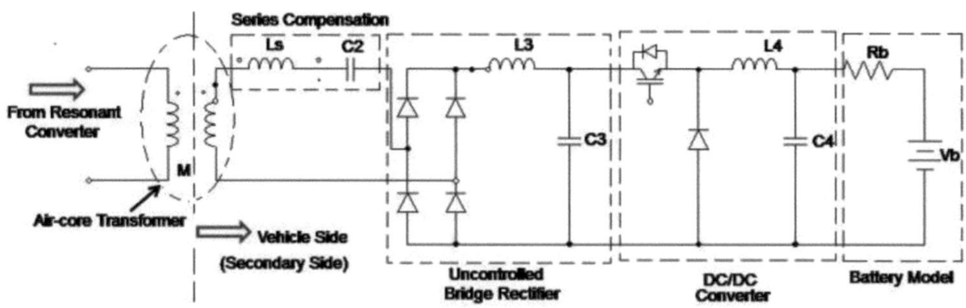

Figure 10.
Components of the electric vehicle section (secondary side).

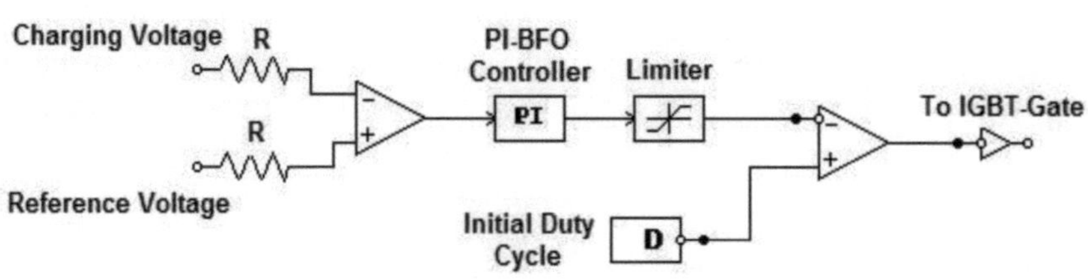

Figure 11.
The control circuit of Buck converter.

4.2.2 Li-ion battery

Charging:

$$V_{bat} = E_0 - R.i - K\frac{Qi^*}{(it - 0.1\,Q)} - K\frac{Qit}{(\,Q - it)} + Ae^{-B.it} \tag{14}$$

Discharge:

$$V_{bat} = E_0 - R.i - K\frac{Q(it + i^*)}{(\,Q - it)} + Ae^{-B.it} \tag{15}$$

4.2.3 NiMH and NiCd battery

Charging:

$$V_{bat} = E_0 - R.i - K\frac{Qi^*}{(|it| - 0.1\,Q)} - K\frac{Q.it}{(\,Q - it)} + e^t \tag{16}$$

$$\text{Discharge}: V_{bat} = E_0 - R.i - K\frac{Q(it + i^*)}{(\,Q - it)} + e^t \tag{17}$$

where V_{bat} is the voltage of the battery (V), E_0 is the battery constant voltage battery (V), K is the constant of polarization ($V = Ah$) or resistance of polarization (Ω), Q is the capacitance of the battery (Ah), $it = \int idt$ is the actual charge of the battery (Ah), R is the internal resistance of the battery (Ω), i is the current of the battery (A), i^* is the current of battery filtered (A), A is the amplitude of exponential zone (V), and B is the time-constant inverse of the exponential zone $(Ah)^{-1}$.

5. Simulation results and discussion

A Matlab/Simulink and m-file general software package are introduced for designing 100-kW EV-battery charger of a renewable-energized public station. The data obtained in Section 3 are used to build the PV arrays. The given parameters from **Table 1** are also used to simulate the ACT transformer needed for the ICPT. The ACT HV windings are proposed to be made from the twisted Litz conductors to minimize the eddy losses [1].

To test the proposed charging system, four 19-Kwh identical EV Honda batteries are charged at the same time. The charging voltage rate is 200 V. When the state of charging (SOC) reaches to 30%, the battery starts to charge. The proposed simulated data for the IGBT-rated voltage and current are 1500 V and 200 A, respectively. To test the robustness of the system against PV fluctuation, the PV-irradiance profile is assumed to vary in the range of 1000 and 250 W/m^2 as shown in **Figure 12a**. Dependently, the PV output terminal voltages with and without using the proposed smart controller are illustrated in **Figures 12b** and **e**, respectively.

The duty ratio of the converter with time according to the variation of irradiance and terminal voltage is shown in **Figure 12c**. The power profile with irradiance variation is clearly illustrated in **Figure 12d**. Through time interval 0.5–0.7, the irradiance reduced from 1000 to 250, and dependently, the power also reduced from 100 kW to less than 20 kW.

This illustrates the importance of the on-grid technique for the station. The modulation-index profile of the three-level interconnected inverter is illustrated in

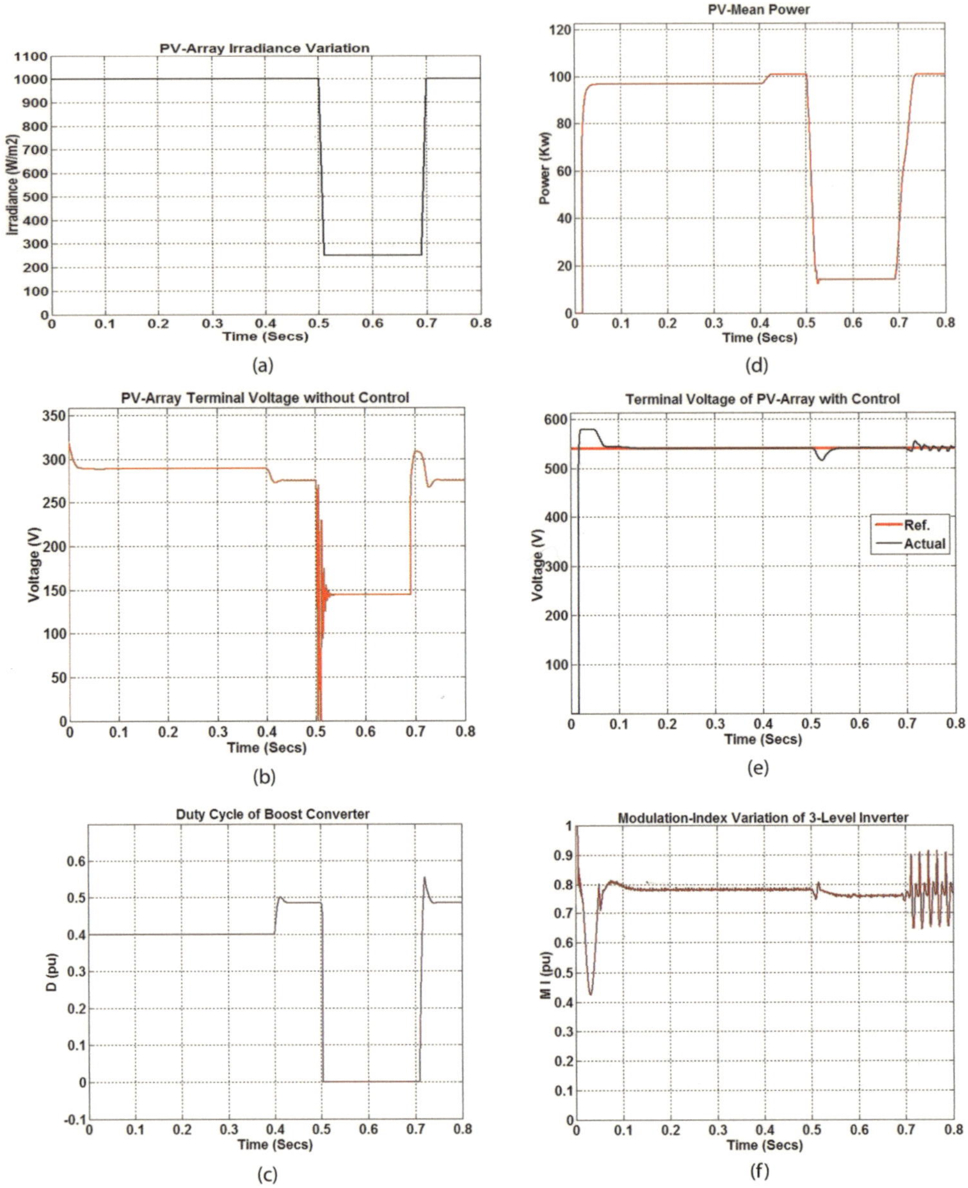

Figure 12.
(a) PV-array irradiance. (b) PV-array terminal voltage profile. (c) Duty cycle for boost converter. (d) PV mean power. (e) PV-array terminal voltage (reference and actual). (f) Modulation-index variation of three-level inverter.

Figure 12f. The figures illustrate the inverter output-voltage changes with the variation of PV irradiance.

Dependently, **Figure 13** shows a comparison for both the 3-level line to line output voltage of the inverter and the grid. It can be noticed that the line voltage maximum value of both is about 540 V ($380\sqrt{2}$). **Figure 14** illustrates the resonant converter output voltage with its high-frequency value that is equals to 30 KHz.

So, both **Figures 15** and **16** describe the high-frequency input voltage and current of the ACT-primary winding, respectively. On the other hand, the voltage and currents of the secondary winding of ACT are illustrated in both **Figures 17** and **18**, respectively.

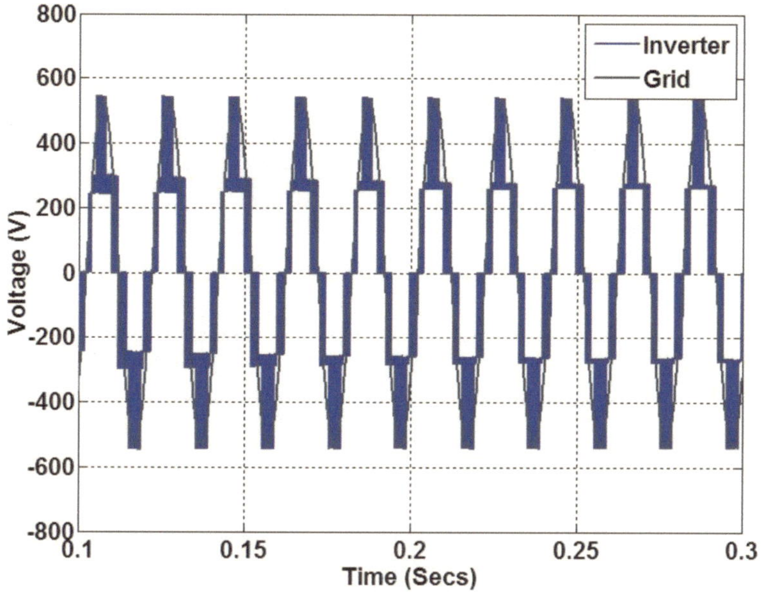

Figure 13.
Line voltage of both three-level inverter and the main grid.

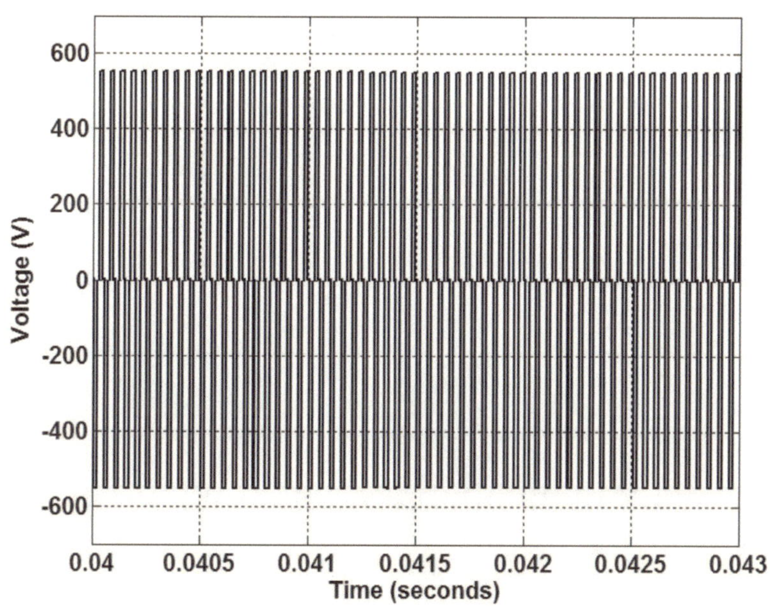

Figure 14.
High-frequency inverter-output voltage.

By using ACT with its turns ratio equal (10:6), when the primary voltage equals to 1000 V, the secondary-winding output voltage equals to 600 V. **Figure 19** describes the simulation results for the initial battery state of charge (SOC) that equals to 30%.

The EV-Honda battery-charging current profile with time is shown in **Figure 20**. According to switching process of the buck converter, the current varies according to the signals of current controller.

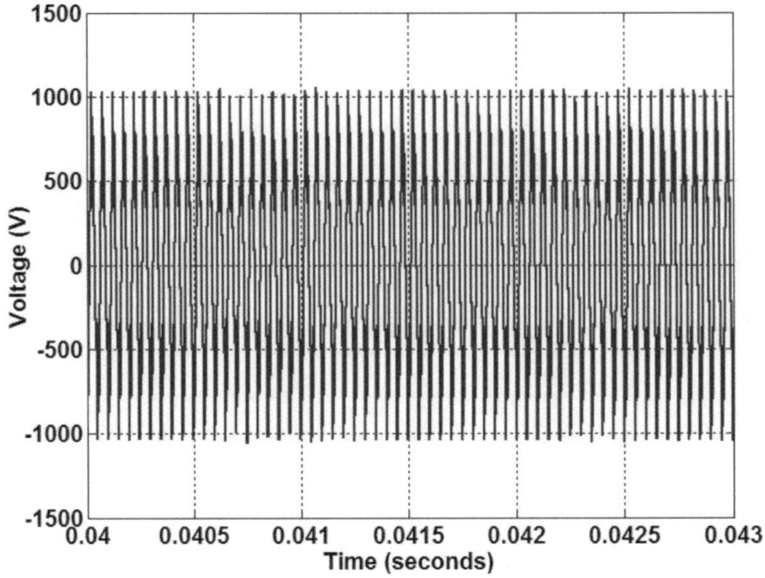

Figure 15.
ACT-primary voltage.

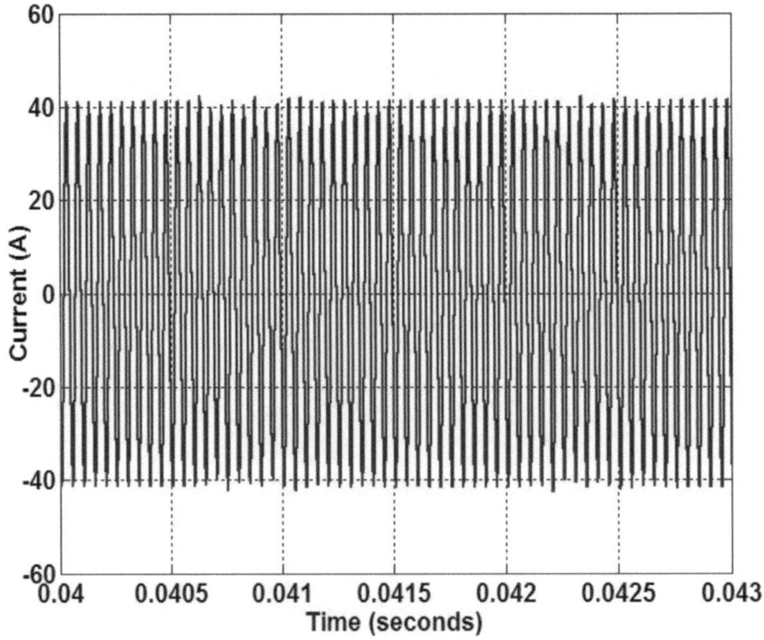

Figure 16.
ACT-primary current.

Also, the voltage level through battery-charging process is demonstrated in **Figure 21**. The value of the battery voltage is controlled to be kept constant, where the charging process is achieved at constant voltage.

However, the overall average charging efficiency can be calculated as follows:

$$\eta_{ol} = \left(\frac{P_o}{P_i} \times 100\right) = \left(\frac{n v_b i_b}{P_m} \times 100\right) = \left(\frac{4 \times 200 \times 90}{94000}\right) \cong 76.6\% \qquad (18)$$

where v_b and i_b are the voltage and current (V,A) of the battery charging, respectively; n is the number of batteries to be charged; and P_m is the PV average power (W).

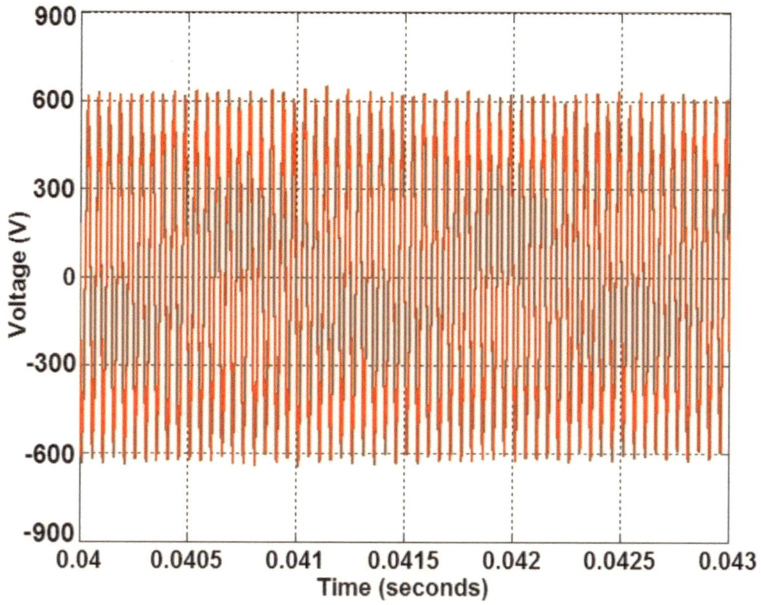

Figure 17.
ACT-secondary voltage.

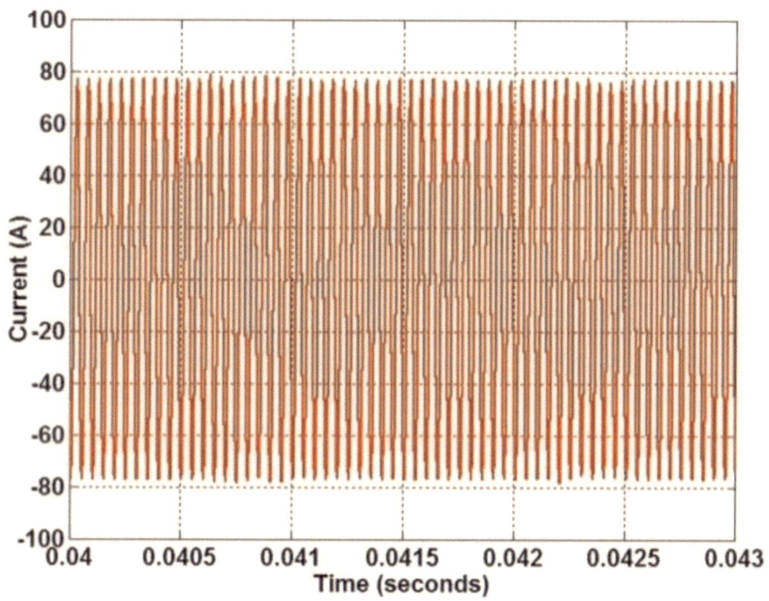

Figure 18.
ACT-secondary current.

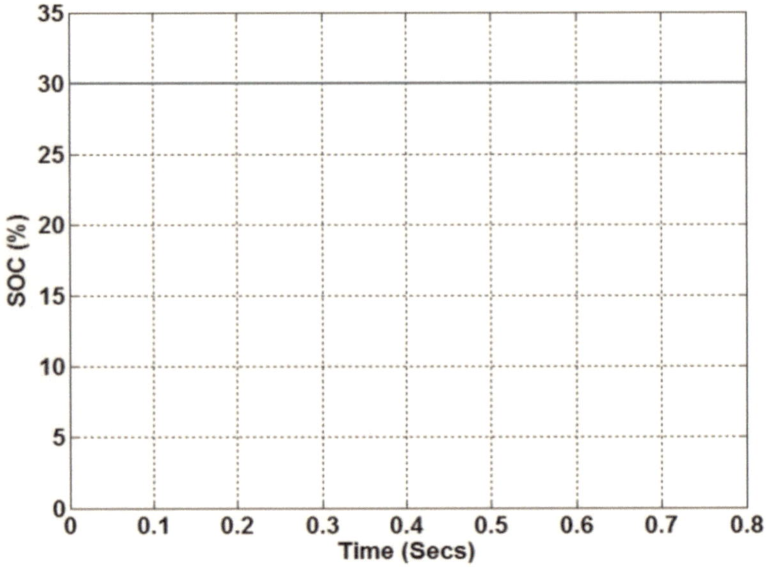

Figure 19.
Initial-battery state of charge (SOC).

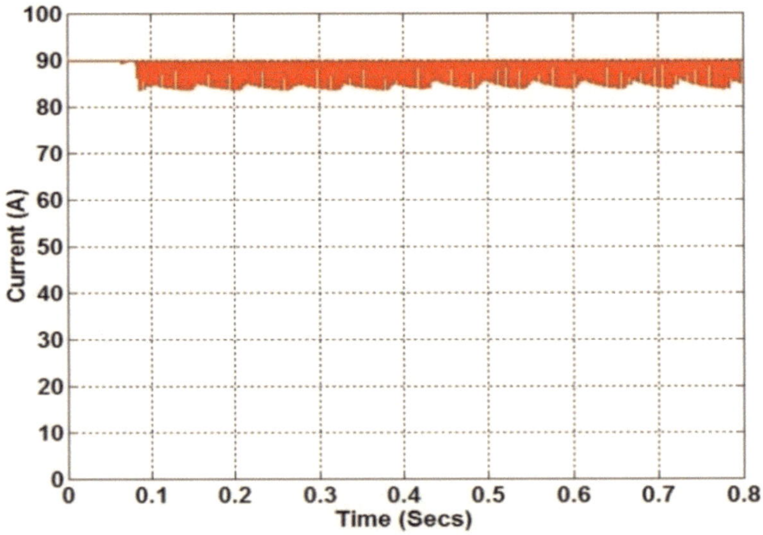

Figure 20.
Battery-charging current.

It should be noted that the efficiency calculated above is the approximate value because of using ideal models for most components used in the design. However, this introduced software package can help researchers and designers to model a solar battery-charging public station. Also, the visibility study for this station does not included her because of several reasons such as the cost of the PV panel is decreased with time and construction cost with other power electronic components differ from place to another in the world.

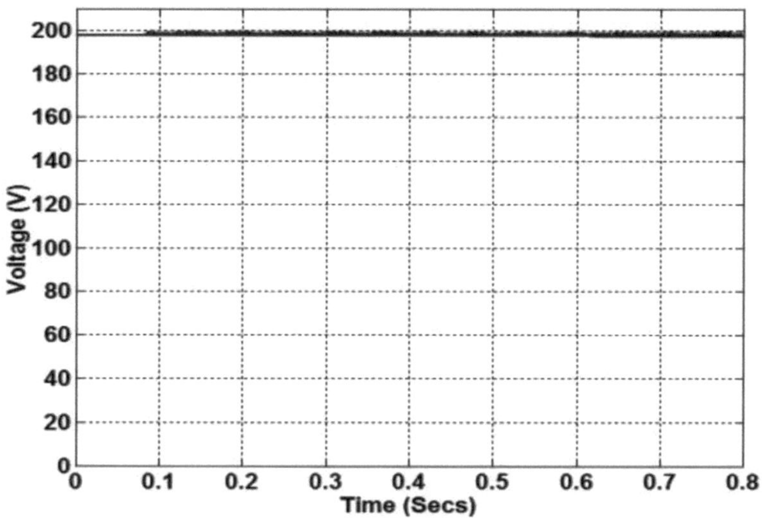

Figure 21.
Battery-terminal voltage profile through charging.

6. Summary and recommendations

The chapter of this book puts in the hands of the reader the main tools required to model and design a solar public station for EV-battery charging. This station is energized from the on-grid PV system and proposed as an off-board 100 kW rating. The proposed system includes a sophisticated un-plugged technique to overcome the problems of wiring and hazards. This technique depends on the inductively coupled power transfer through the air-core transformer with a large air gap. Two main parts are implied and classified according to the winding of the transformer: primary and secondary parts. The primary part is placed on the floor of the station, and the other part is installed at the bottom of the car. A Matlab/Simulink with m-file is powerful tool that used to model and simulate all components such as PV model, maximum power point tracking (MPPT) technique, boost inverter, three-level inverter, series-series (SS) compensation with high-frequency resonant converter, air-core transformer, rectifier, switched-mode power converter, and battery-charging smart controller. The proposed station is tested by charging four 19-kW batteries for EV from Honda Company. Simulation results include the battery state of charge (SOC), charging voltage, and currents. Using high-frequency compensation with switched-mode power converter and smart PI-tuning controller for the charger make the charging process is so fast. The ideal overall efficiency for the proposed solar station is relatively high due to assuming ideal cases for all modeled components. Finally, due to the rapid change in the cost of all components, especially the PV panels, this chapter does not imply a visible study for the station. So, it is preferable to conduct a visible study for each country separately.

Author details

Essamudin Ali Ebrahim
Power Electronics and Energy Conversion Department, Electronics Research Institute, Cairo, Egypt

*Address all correspondence to: essamudin@eri.sci.eg; essamudin@yahoo.com

References

[1] Ebrahim EA. A general software package for modelling a contact-less electric-vehicles battery-charging public-station fed from on-grid photo-voltaic array. Elsevier Journal of Electrical Systems and Information Technology. 2018;**5**:271-286. DOI: 10.1016/j.jesit.2018.04.001

[2] Wang C-S, Stielau OH, Covic GA. Design considerations for a contactless electric vehicle battery charger. IEEE Transactions on Industrial Electronics. 2005;**52**(5):1308-1314. DOI: 10.1109/TIE.2005.855672

[3] Sallán J, Villa JL, Llombart A, Fco J. Sanz: Optimal design of ICPT systems applied to electric vehicle battery charge. IEEE Transactions on Industrial Electronics. 2009;**56**(6):2140-2149. DOI: 10.1109/TIE.2009.2015359

[4] Villa JL, Sallán J, Llombart A, Fco Sanz J. Design of a high frequency inductively coupled power transfer system for electric vehicle battery charge. Elsevier Journal of Applied Energy. 2009;**86**:355-363. DOI: 10.1016/j.apenergy.2008.05.009

[5] Tabari M, Yazdani A. Stability of a dc distribution system for power system integration of plug-In hybrid electric vehicles. IEEE Transactions on Smart Grid. 2014;**5**(5):2464-2674. DOI: 10.1109/TSG.2014.2331558

[6] Kim S, Kang F-S. Multifunctional on-board battery charger for plug-in electric vehicles. IEEE Transactions on Industrial Electronics. 2015;**62**(6): 3460-3472. DOI: 10.1109/TIE.2014.2376878

[7] Janghorban S, Teixeira C, Holmes DG, McGoldrick P, Yu X. Magnetics design for a 2.5-kW battery charger. In: Proceedings of Australasian Universities Power Engineering Conference, AUPEC 2014, Curtin University, Perth, Australia. 2014. pp. 1-6

[8] Ebrahim EA. A novel approach of bacteria-foraging optimized controller for DC motor and centrifugal pump set fed from photo-voltaic Array. Journal of Next Generation Information Technology. 2015;**6**(1):21-31

[9] Miskovski D, Williamson SS. Modelling and simulation of a photovoltaic (PV) based inductive power transfer electric vehicle public charging station. In: Proceedings of IEEE Transportation & Electrification Conference and Expo (ITEC 2013). 2013. pp. 1-6. DOI: 10.1109/ITEC.2013.6573491

[10] Halder T. Charge controller of solar photo-voltaic panel fed (SPV) battery. In: Proceedings of India Inter. Conf. on Power Electronics (IICPE), New Delhi, India. 2011. pp. 1-4. DOI: 10.1109/IICPE.2011.5728056

[11] Robalino DM et al. Design of a docking station for solar charged electric and fuel cell vehicles. In: Proceedings of 2009 International Conference on Clean Electrical Power, Capri. 2009. pp. 655-660

[12] Giroux P, Sybille G, Osorio C, Chandrachood S. Grid-Connected PV Array. USA: Research Centre Press, Mathwork Co.; 2012

[13] Mathwork Corporation: Matlab/Simulink 2012b, User's Guide. USA; 2012

[14] Available from: http://www.solardesigntool.com/components/module-panel-olar/Sunpower/514/SPR-305-WHT-U/specification-data-sheet.html

[15] Esram T, Chapman PL. Comparison of photovoltaic array maximum power point tracking techniques. IEEE

Transactions on Energy Conversion. 2007;**22**(2):439-449. DOI: 10.1109/TEC.2006.874230

[16] Harrabi N, Soussi M, Aitouche A. Maximum power control for photovoltaic power based on fuzzy Takagi-Sugeno model. Journal of Electrical Engineering. 2015;**25**:1-11

[17] Pouresmaeil E, Miracle DM, Bellmunt OG. Control scheme of three-level NPC inverter for integration of renewable energy resources into AC grid. IEEE Systems Journal. 2012;**6**(2):242-253

[18] Subsingha W. Design and analysis three phase three level diode-clamped grid connected inverter. Energy Procedia. 2016;**89**:130-136

[19] Powerex Company Catalogue: Application Notes to Three-Level Inverter Technology, First Release; 2009:1-12

[20] Tremblay O, Dessaint L-A. Experimental validation of a battery dynamic model for EV applications. World Electric Vehicle Journal. 2009;**3**

[21] El Shahat A, Haddad R, Kalaani Y. Lead acid battery modeling for photo-voltaic applications. Journal of Electrical Engineering. 2015;**15**

Section 3

Advances in Energy Storage Systems

Chapter 5

SiO_x as a Potential Anode Material for Li-Ion Batteries: Role of Carbon Coating, Doping, and Structural Modifications

Hyeon-Woo Yang and Sun-Jae Kim

Abstract

Despite the high energy density of SiO_x, its practical use as an anode material for Li-ion batteries is hindered by its low electronic conductivity and sluggish electron transport kinetics. These disadvantageous properties result from the insulating nature of SiO_2, which leads to electrical contact loss and poor cyclability. Herein, we synthesized a C-SiO_x composite based on amorphous carbon and a SiO_x matrix via the alcoholysis reaction between $SiCl_4$ and ethylene glycol. We then used nonpolar benzene to simultaneously achieve homogenous dispersion of the Si source and the formation of a carbon coating layer, resulting in the formation of a (C-SiO_x)@C composite with exceptional electrochemical properties. Next, we performed structural modifications using Ti doping and a multiple-carbon matrix to successfully fabricate a (C-$Ti_xSi_{1\_}xO_y$)@C composite. The combination of Ti doping and carbon coating greatly enhanced the conductivity of SiO_x; moreover, the incorporated carbon acted as an effective oxide buffer, preventing structural degradation. The (C-$Ti_xSi_{1\_}xO_y$)@C composite exhibited excellent capacity retention of 88.9% over 600 cycles at 1 A g^{-1} with a capacity of 828 mAh g^{-1}.

Keywords: lithium ion battery, SiO_x anode, multiple carbon matrix, doping

1. Introduction

Silicon (Si) is a key anode material for fabricating next-generation Li-ion batteries (LIBs) with longer cycle life and higher energy density to help meet the growing market demand for electric vehicles (EVs) and hybrid cars [1–3]. As a host material for lithium, Si is earth-abundant and delivers a high theoretical capacity of 3578 mAh g^{-1} (compared with 372 mAh g^{-1} for carbon-based electrodes) [4–6]. Nevertheless, the large volumetric expansion (~400%) of Si anodes results in degradation of Si particles and destruction of the solid-electrolyte interphase (SEI) [7–9]. These issues can induce drastic capacity fade and even overall damage to the electrodes, thereby hindering the commercial application of Si anodes in LIBs.

Silicon suboxide (SiO_x, $0 < x < 2$) has attracted considerable interest as a potential alternative to Si because of its enhanced cycling stability. SiO_x not only exhibits a relatively small volume expansion but also forms Li_2O and Li silicates that serve as buffer media for Si during the first lithiation process [10–12]. As a result, SiO_x

exhibits better cycling performance than Si. Nevertheless, the low electronic conductivity and sluggish electron transport kinetics of SiO_x resulting from the insulating property of SiO_2 lead to poor electrochemical performance and have hindered the application of SiO_x as anode materials for commercialized LIBs [13–16]. Many researchers have proposed strategies to address these issues, resulting in progress such as the development of carbon-coated SiO_x composites. Although the improved electrical conductivity achieved by carbon coating can improve the electrochemical performance of SiO_x, complicated, multi-step, and high-temperature processes are required [17–20]. For instance, Liu et al. developed a Si-Void@SiO_x nanowire composite using thermal evaporation/chemical etching of a mixed powder of SiO and ZnS at high temperatures (1250 and 1650°C) [18]. In addition, Han et al. prepared a SiO@C composite using a two-step process with SiO powder as the raw material; after ball milling for 3 h at 3000 rpm, the ball-milled SiO particles were calcined at 700°C using sodium dodecylbenzene sulfonate [19].

To avoid the complicated and costly processes adopted in previous studies, in this study, a simple and cost-effective one-pot synthesis method was developed to fabricate a carbon-incorporated/carbon-coated SiO_x ((C-SiO_x)@C) composite. We attempted to simultaneously form interconnected carbon paths in the composite and encapsulate the surface with carbon using ethylene glycol and benzene. We further attempted to fabricate a SiO_x composite with superior electrochemical performance by maximizing the electrical conductivity through Ti doping. Ti doping can result in the formation of TiSi alloys, which are beneficial for improving the cyclic stability of LIB electrode materials [21, 22]. In addition, black TiO_{2-x} has been reported to exhibit higher conductivity than pristine white TiO_2 because of the existence of Ti^{3+} (corresponding to an oxygen deficiency) in the structure. Thus, we suspected that Ti^{3+} doping might lead to outstanding electrochemical performance [23–25].

In the current study, we prepared a Ti^{3+}-doped and carbon-incorporated/carbon-coated SiO_x ((C-$Ti_xSi_{1-x}O_y$)@C) composite and investigated the effects of these structural modifications. The electrochemical performance of the (C-$Ti_xSi_{1-x}O_y$)@C composite was greatly improved compared with that of carbon-incorporated SiO_x (C-SiO_x) and a (C-SiO_x)@C composite. The electrochemical performance of the (C-$Ti_xSi_{1-x}O_y$)@C composite was greatly improved compared with that of the C-SiO_x and (C-SiO_x)@C composite. The initial discharge capacity of the (C-$Ti_xSi_{1-x}O_y$)@C composite at 0.1 A g^{-1} was ~1304 mAh g^{-1}, which was ~4 times higher than that of C-SiO_x under the same conditions. Furthermore, the (C-$Ti_xSi_{1-x}O_y$)@C composite delivered a capacity retention of ~88.9% over 600 cycles at a higher current density of 1 A g^{-1} with a high coulombic efficiency of ~99%.

2. Experimental

Preparation of C-SiO_x: First, 13 mL of ethylene glycol (EG, 99.9%, Samchun Co.) was added to 20 mL of $SiCl_4$ (99%, Wako Co.) under vigorous stirring. The mixture was rapidly transformed into a mineral-like solid, which was converted into C-SiO_x powder by heat treatment at 725°C for 1 h under vacuum.

Preparation of (C-SiO_x)@C: First, 13 mL of EG was poured into a mixture of 20 mL of $SiCl_4$ and 50 mL of benzene (99.5%, Daejung Co.) under vigorous stirring. Benzene was used to control the reaction between EG and $SiCl_4$ necessary for the synthesis of the powder. The mineral-like solid formed through the alcoholysis reaction was transformed into a (C-SiO_x)@C composite by heat treatment at 725°C for 1 h under vacuum.

Preparation of $(C\text{-}Ti_xSi_{1-x}O_y)@C$: First, 50 μL of $TiCl_4$ (99.9% SHOWA) was dissolved in EG (13 mL) by stirring for 1 day. Then, nonpolar $SiCl_4$ (20 mL) was uniformly distributed in 50 mL of benzene for 30 min. After the solution was added to the mixture of EG and $TiCl_4$ under vigorous stirring, the solution-state mixture was converted into a yellow mineral-like solid containing Si, Ti, O, and C. Finally, the $(C\text{-}Ti_xSi_{1-x}O_y)@C$ composite was obtained by heat treatment in a tube furnace at 725°C for 1 h under vacuum.

Materials characterization: The morphologies of all the samples were characterized using field-emission scanning electron microscopy (FESEM; SU-8010 and S-4700, Hitachi Co.), high-resolution transmission electron microscopy (HRTEM; JEM 2100F, JEOL); and Cs-corrected TEM with cold FEG (Cs-TEM, JEM-ARM200F, JEOL). Energy-dispersive X-ray spectroscopy (EDS) coupled with TEM was used for local elemental analyses. The crystal structures were characterized using X-ray diffraction (XRD; Rigaku, D/MAX-2500). The state of carbon was analyzed using Raman spectroscopy (FEX, Nost Co., Ltd.; 532-nm wavelength), and X-ray photoelectron spectroscopy (XPS; K-alpha, Thermo Scientific Inc.) was employed to obtain further information about Si2p, C1s, O1s, Li1s, and F1s. The carbon content of the composite was measured using a carbon/sulfur analyzer (CS-2000, ELTRA GmbH).

Electrochemical characterization: For the electrochemical characterization, all the powders were crushed using a 3D mixer (Turbula mixer, DM-T2, Daemyoung Co.) with 5-mm zirconia balls at 50 rpm for 24 h to achieve a uniform particle distribution. All the samples were first mixed with Super P (SP, TIMCAL, Super P Li) and sodium-carboxymethyl cellulose (Na-CMC, Sigma Aldrich Co.) in an active material/Super P/CMC weight ratio of 70/20/10; deionized (DI) water was added to form a homogeneous slurry. Electrochemical characterization of the electrode was performed using CR2032 coin-type cells, with a lithium metal foil used as the counter electrode. The electrodes were dried in a vacuum oven at 80°C for 24 h before being transferred to an Ar-filled glove box for cell assembly. The electrolyte used was a solution of 1.2 M $LiPF_6$ dissolved in a mixture of ethylene carbonate and dimethyl carbonate (3:7 v/v, Panax Etec) containing 3% vinylene carbonate additive. The coin cells were charged and discharged between 0.01 and 1.5 V (vs. Li/Li^+) by applying various currents ranging from 0.1 to 5 A g^{-1} at 25°C for the electrochemical characterization. Electrochemical impedance spectroscopy (EIS) analysis (Bio-Logic Co., VMP3) was performed in the frequency range of 1 MHz to 1 mHz with an AC amplitude of 10 mV.

3. Synthesis of SiO_x active materials for highly enhanced electrochemical performance

3.1 Amorphous SiO_x and carbon matrix

Images of the SiO_x composites formed via the alcoholysis reaction before and after heat treatment are presented in **Figure 1a**. The alcoholysis mechanism between the silicon precursor, silicon tetrachloride ($SiCl_4$), and ethanol was as follows:

$$SiCl_4 + 4C_2H_5OH \rightarrow Si(OC_2H_5)_4 + 4HCl\uparrow \quad (1)$$

$$Si(OC_2H_5)_4 + \text{Heat treatment}\ (> 600^\circ C) \rightarrow SiO_2 + 2O(C_2H_5)_2 \quad (2)$$

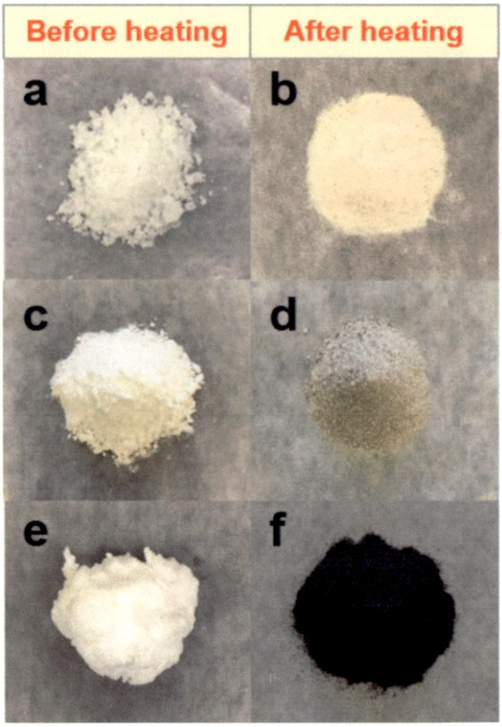

Figure 1.
Image of synthesized SiO_x composite prepared using $SiCl_4$ and water (a and b), $SiCl_4$ and ethanol (c and d), and $SiCl_4$ and ethylene glycol (e and f) respectively.

$$C_2H_5OH \rightarrow C + 3H_2\uparrow + CO \text{ or } CO_2\uparrow \tag{3}$$

During the reaction between $SiCl_4$ and ethanol, tetraethoxysilane (TEOS) was formed (1). The TEOS was converted into SiO_2 phase by the heat treatment (>600°C) (2). Simultaneously, the residual ethanol generated carbon and CO or CO_2 gas (3). As observed in **Figure 1a**, the mineral-like solid transformed into a gray SiO_x powder after heat treatment. In addition, SiO_2 and HCl were formed by the reaction between $SiCl_4$ and water, and a white SiO_2 powder was obtained after the heat treatment.

In contrast to these reactions, the alcoholysis mechanism between $SiCl_4$ and EG was as follows:

$$xSiCl_4 + yC_2H_4(OH)_2 \rightarrow Si_x(O_2C_2H_2)_y + \alpha HCl\uparrow \tag{4}$$

$$Si_x(O_2C_2H_2)_y + \text{Heat treatment}\left(> 600^{\circ}C\right) \rightarrow Si_x(OC)_y + yCO\uparrow + yH_2\uparrow \tag{5}$$

$Si_x(O_2C_2H_2)_y$ was formed by the alcoholysis reaction between $SiCl_4$ and EG (4). The obtained $Si_x(O_2C_2H_2)_y$ was transformed into the completely black $Si_x(OC)_y$ after heat treatment (5). The $Si_x(OC)_y$ powder composed of a carbon and SiO_x matrix was labeled as the carbon-incorporated SiO_x (C-SiO_x) composite.

We used several characterization techniques to confirm the carbon-based complex of the C-SiO_x composite and the mechanism proposed above. The XRD pattern in **Figure 2a** reveals broad peaks over the range of 10–30°, which can be indexed to the amorphous phase of the C-SiO_x composite. In addition, the elemental bonding properties of the C-SiO_x composite were investigated using XPS analyses.

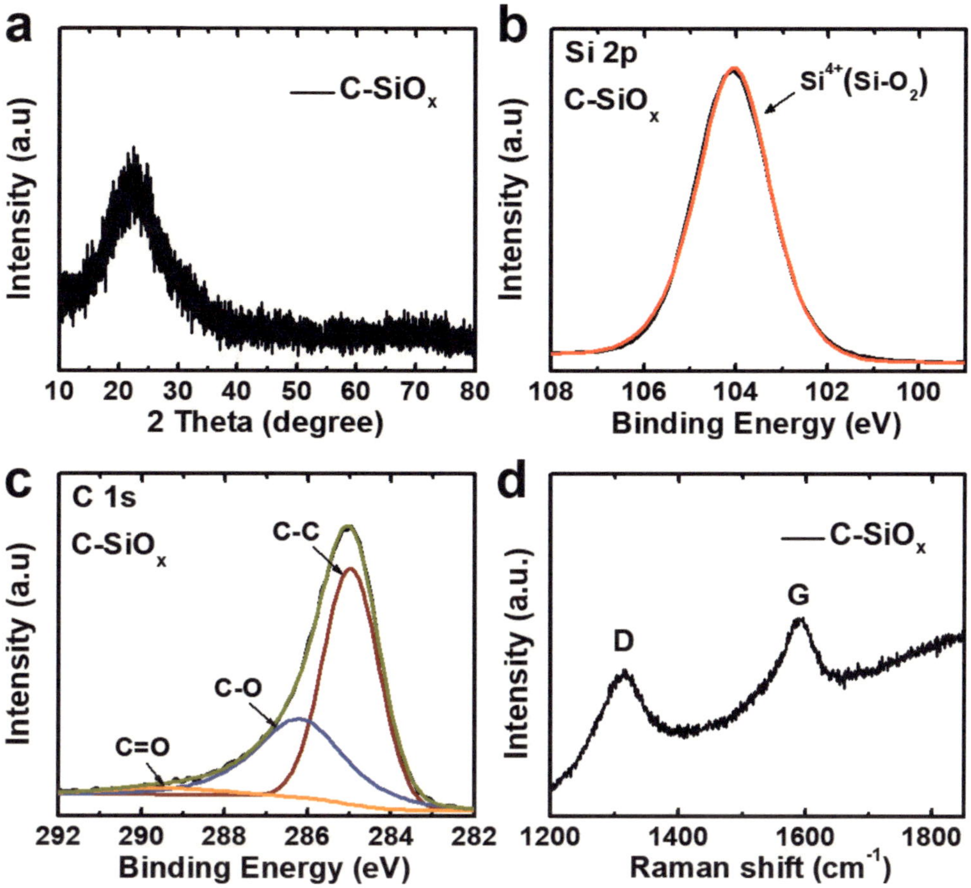

Figure 2.
(a) XRD patterns, XPS spectra showing (b) Si 2p peak and (c) C 1s peak, and (d) Raman spectra of C-SiO_x composite.

As observed in **Figure 2b**, the characteristic peaks of amorphous SiO_2 were detected at 103.8 eV in the Si 2p spectra, whereas those for Si–Si bonding were not observed, indicating that the Si particles were completely surrounded by SiO_2. **Figure 2c** shows that various carbon-related bonds, such as C—C (285 eV), C—O (286.7 eV), and C=O (298.4 eV), were detected in the C 1s spectra, confirming the existence of the carbon-based complex in the C-SiO_x composite. The XPS analysis confirmed that the C-SiO_x composite was composed of Si^{4+} and C—C, C—O, and C=O bonds. The carbon matrix in the C-SiO_x composite was further characterized using Raman spectroscopy, and the results are presented in **Figure 2d**. After carbonization was achieved by the heat treatment, strong peaks centered at 1360 and 1580 cm^{-1} appeared in the Raman spectra of the C-SiO_x composite, corresponding to the disordered carbon band (D band) and graphitic carbon band (G band), respectively.

Figure 3 presents the cycle performance profiles of the SiO_x composites synthesized at a current density of 0.1 A g^{-1} using water, ethanol, and EG, respectively. The C-SiO_x composite exhibited a first discharge capacity of 330 mAh g^{-1} with great reversibility, whereas the SiO_x composites prepared using water and ethanol delivered first discharge capacities of 31 and 60 mAh g^{-1}, respectively. Although the SiO_x structure resulted in a low reversible capacity, the low electrical conductivity of SiO_2 was overcome by synthesizing the C-SiO_x composite with a carbon matrix. In addition, the SiO_x structure was advantageous for achieving good cyclability because the presence of SiO_2 buffers effectively reduced the large volume change of

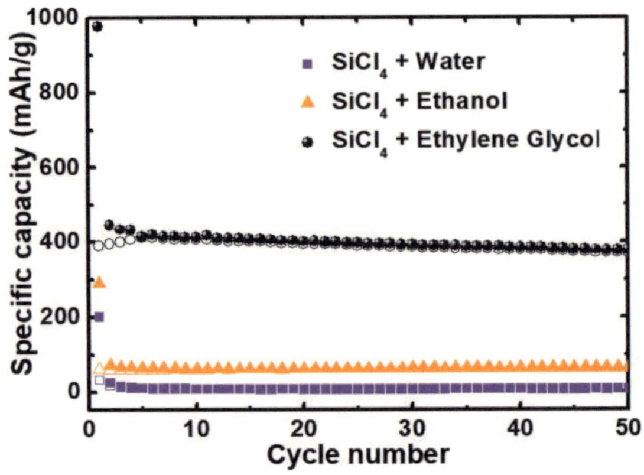

Figure 3.
Cyclic performance of synthesized SiO_x composite prepared using water, ethanol, and ethylene glycol.

Si during charge/discharge cycles. Therefore, the electrochemical tests revealed the excellent electrochemical performance of the C-SiO_x composite.

3.2 Effect of carbon coating on the surface of SiO_x particles

To improve the poor reversible capacity resulting from the low electrical conductivity, we next designed a novel carbon-incorporated/carbon-coated SiO_x ((C-SiO_x)@C) composite using EG with benzene to achieve the homogenous distribution of the Si source and simultaneous formation of a multiple-carbon matrix in the composite. In particular, the use of nonpolar benzene enabled the formation of uniformly disperse nonpolar $SiCl_4$ via dispersion forces, which potentially minimized the aggregation of Si nanoparticles and contributed to the formation of a carbon framework in the SiO_x composite. Moreover, the conductive carbon was completely coated on the surface of each SiO_x particle, which not only provided a fast electron transport path but also effectively prevented structural failure resulting from the large volume expansion during charge/discharge, which led to great enhancement of the electrochemical properties of the SiO_x composite.

To verify the successful preparation of the (C-SiO_x)@C composite and our above hypotheses, we performed several experiments. Structural variation using benzene was identified using XPS analyses, which confirmed the chemical states of each element in the composite. As observed in **Figure 4a** and **b**, the (C-SiO_x)@C composite was based on SiO_2, similar to the C-SiO_x composite. However, the higher intensity of the C—C bond compared with that of the C—O bond in the C 1s spectra of C for the (C-SiO_x)@ composite indicates the presence of a multiple-carbon matrix derived from EG and benzene. A pitch-coated C-SiO_x composite was also prepared using ~11 wt% pitch carbon to confirm the effects of the benzene-based carbon coating, as the C content in the (C-SiO_x)@C composite measured using a carbon/sulfur determinator was estimated to be 11 wt%.

In the XRD patterns of the obtained powders of the C-SiO_x, pitch-coated C-SiO_x and (C-SiO_x)@C composite, only a broad peak at approximately 25° was observed without the appearance of a crystalline Si peak (**Figure 5a**). The Raman spectra of the samples revealed two strong peaks at 1360 and 1580 cm^{-1}, which were assigned to the D band and G band from the carbon, respectively, as observed in **Figure 5b**. The intensity ratio between the D and G bands indicated the crystallinity of graphitic carbon. In contrast, much weaker peaks were observed for the C-SiO_x and pitch-coated C-SiO_x, suggesting the presence of a multiple-carbon matrix composed of highly graphitic carbon in the (C-SiO_x)@C composite.

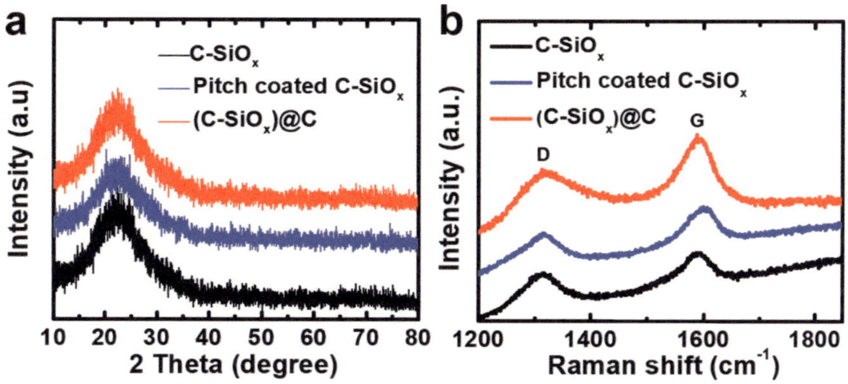

Figure 4.
XPS spectra showing (a) Si 2p peak and (b) C 1s peak of (C-SiO$_x$)@C composite.

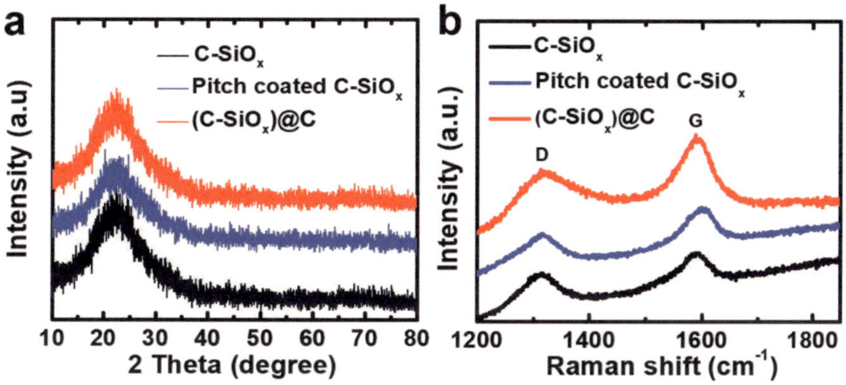

Figure 5.
(a) XRD patterns and (b) Raman spectra of C-SiO$_x$, pitch coated C-SiO$_x$, and (C-SiO$_x$)@C composite.

To further evaluate the carbon-based complex of the (C-SiO$_x$)@C composite in detail, we examined the morphology of (C-SiO$_x$)@C using TEM and compared it with those of the C-SiO$_x$ and pitch-coated C-SiO$_x$. As observed in **Figure 6a–c**, TEM analysis confirmed that an amorphous matrix surrounded the uniformly dispersed Si nanoparticles (3–4 nm size); the amorphous layer was also homogeneously coated on the surface of all the particles of the (C-SiO$_x$)@C composite. TEM mapping analysis indicated that the outer coating layer was mainly composed of C, whereas the core was composed of C, Si, and O from the interconnected structures consisting of amorphous phases such as SiO_2 and C (**Figure 6d**). These analyses indicated that the interconnected carbon paths and homogenous carbon coating were successfully prepared, resulting in considerable improvement of the electrochemical performance of the novel (C-SiO$_x$)@C composite. However, the crystalline Si nanoparticles and carbon coating layer were not observed in the TEM images of the C-SiO$_x$, whereas for the pitch-coated C-SiO$_x$, a pitch carbon layer coated on amorphous SiO$_x$ and a C matrix are observed in **Figure 6e** and **f**, respectively.

The electrochemical behavior of the (C-SiO$_x$)@C composite was investigated using galvanostatic measurements in Li cells and was compared with that of C-SiO$_x$ and pitch-coated C-SiO$_x$. **Figure 7a** presents the charge/discharge capacities of the samples at a current density of 0.1 A g^{-1} in the voltage range of 0.01–1.5 V. The (C-SiO$_x$)@C electrode delivered a high initial discharge capacity (925 mAh g^{-1}) and high stability during repeated charge/discharge cycles. Notably, up to ~92% of the initial discharge capacity was maintained, whereas the capacity of the pitch-coated

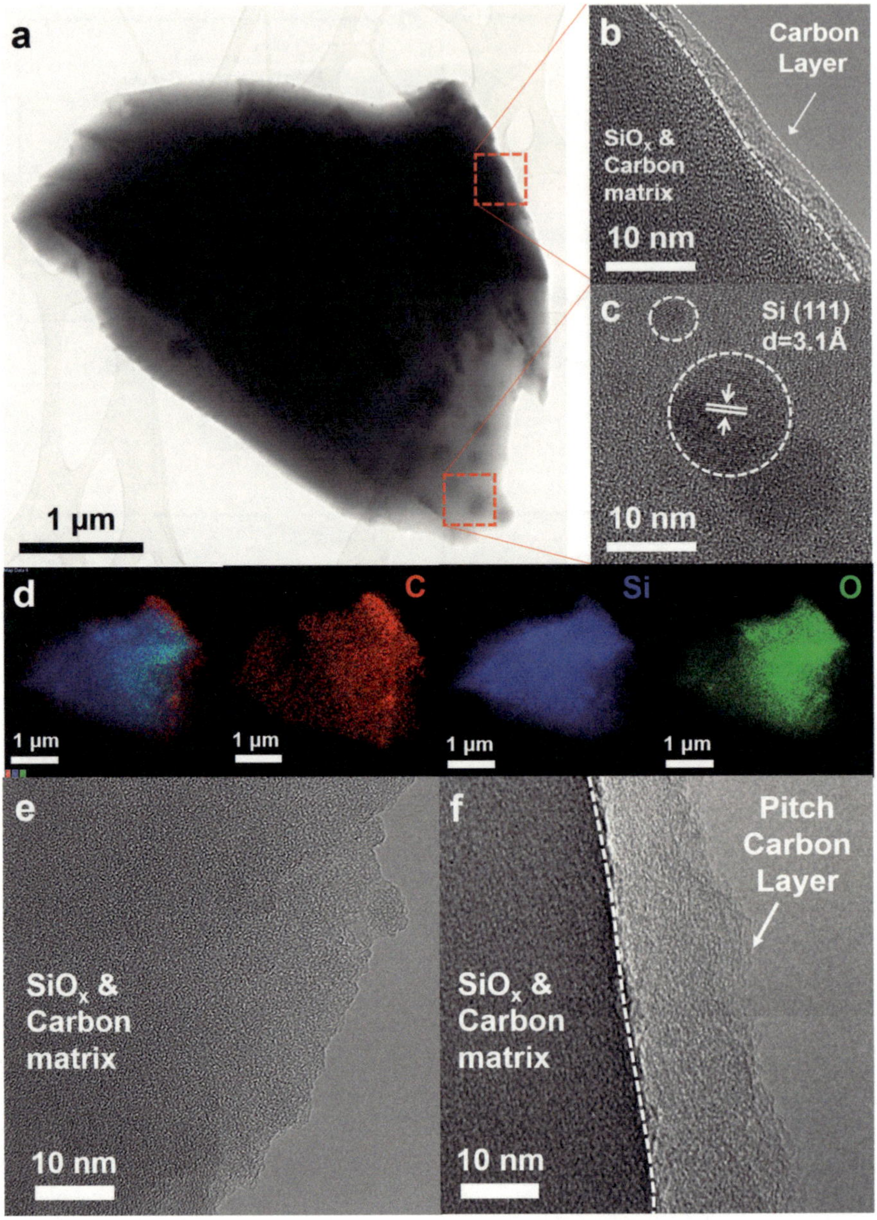

Figure 6.
(a)–(c) TEM images of (C-SiO$_x$)@C composite and (d) TEM elemental mapping images of C, Si, and O in the (C-SiO$_x$)@C composite. TEM images of (e) C-SiO$_x$ and (f) pitch-coated C-SiO$_x$.

SiO$_x$ gradually decreased under the same conditions. The pitch-coated SiO$_x$ electrode exhibited a high initial charge/discharge capacity of 1891/786 mAh g^{-1} but only 88.8% retention of its initial capacity over 100 cycles. In addition, the first discharge capacity of the pristine SiO$_x$ was only ~400 mAh g^{-1}, which is less than half of that of the (C-SiO$_x$)@C composite. EIS characterization was performed for the C-SiO$_x$, pitch-coated C-SiO$_x$, and (C-SiO$_x$)@C electrodes before cycling, and the electrical conductivity was substantially improved by carbon coating. As observed in **Figure 7b**, the (C-SiO$_x$)@C composite exhibited a smaller charge-transfer resistance than the C-SiO$_x$ and pitch-coated C-SiO$_x$, indicating that the carbon framework derived from EG and benzene enhanced the electrical conductivity of

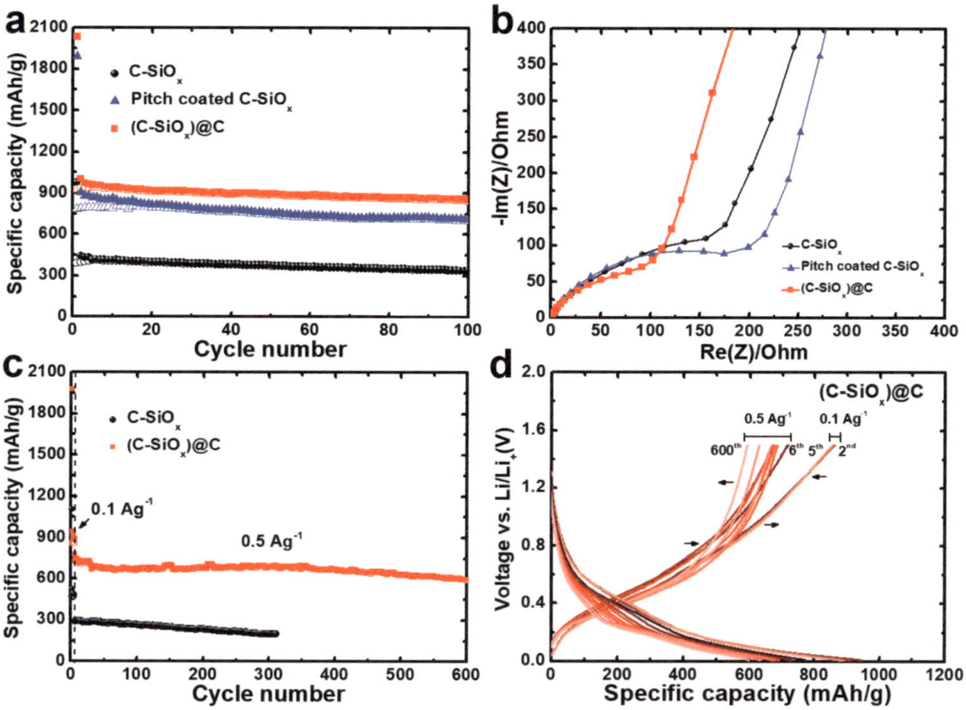

Figure 7.
Electrochemical characteristics of C-SiO$_x$, pitch-coated SiO$_x$, and (C-SiO$_x$)@C composite. (a) Cycling performance at a current density of 0.1 A g^{-1} and (b) Nyquist plots in the frequency range of 1 MHz to 1 mHz with an AC amplitude of 10 mV. (c) Cycling performance of C-SiO$_x$ and (C-SiO$_x$)@C composite at a current density of 0.5 A g^{-1}. (d) Charge/discharge curves of (C-SiO$_x$)@C composites over 600 cycles.

the SiO$_x$ particles. Additionally, the (C-SiO$_x$)@C composite exhibited satisfactory performance at a higher current density. As observed in **Figure 7c** and **d**, 591 mAh g^{-1} of the discharge capacity of the (C-SiO$_x$)@C composite at 0.5 A g^{-1} was maintained over 600 cycles with a coulombic efficiency of ~99%, whereas the capacity of C-SiO$_x$ drastically declined under the same conditions. These results indicate that the multiple-carbon matrix in the (C-SiO$_x$)@C composite not only provided high electrical conductivity but also prevented the severe structural degradation that generally accompanies the large volume change during charge/discharge.

The structural differences between the C-SiO$_x$ and (C-SiO$_x$)@C composite electrodes were clearly determined using XPS analyses. The chemical state of each element in the compound was identified, as shown in **Figure 8**. For the Si 2p spectra, the electrodes exhibited peaks at 101.6 and 103.5 eV, corresponding to lithium silicates (Li_xSiO_y) and Si—O bonding, respectively. These lithium silicate phases are known irreversible products formed during the first cycle, and their detection in the SEI layer of Si-based electrodes has been previously reported [26, 27]. The presence of the higher lithium silicates peak indicates that the rate of irreversible consumption was higher than that in the C-SiO$_x$ electrode. For the C 1s spectrum of the C-SiO$_x$ and (C-SiO$_x$)@C electrodes, the peak at 285.0 eV was assigned to the C—C bonds in the carbon-based complex of the SiO$_x$ anode. The peak at 287.0 eV corresponding to C—O originates from the SiO$_x$ composite and CMC binder. For the C-SiO$_x$ electrode, the absence of the peak at 289.1 eV indicates the disappearance of C=O bonding. Instead, a new peak appeared at 290.3 eV, which is attributed to the formation of SEI layer components such as lithium carbonate (Li_2CO_3) and lithium alkyl carbonates [28]. For the Li 1s spectra, the peaks at 54.5, 56.0, and 56.8 eV are assigned to the formation of SEI layer components such as Li_2O, lithium

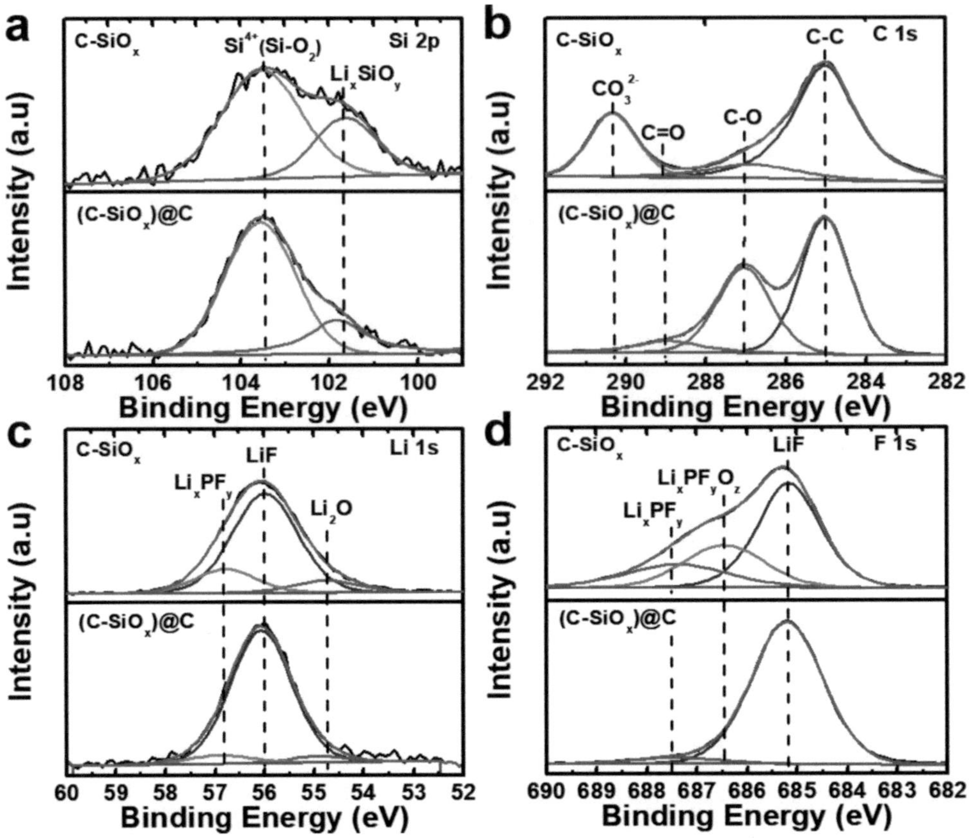

Figure 8.
XPS spectra of surface of C-SiO$_x$ and (C-SiO$_x$)@C electrodes in the first cycle: (a) Si 2p, (b) C 1s, (c) Li 1s, and (d) F 1s branches.

fluoride (LiF), and Li_xPF_y, respectively. Compared with the C-SiO$_x$ electrode, weak Li_2O and Li_xPF_y peaks and a strong LiF peak were observed for the (C-SiO$_x$)@C electrode, indicating that LiF was the main component of the SEI layer. The F 1s spectra for the (C-SiO$_x$)@C anode contained a very strong peak at 685.2 eV attributable to LiF in addition to very weak peaks assigned to $Li_xPF_yO_z$ (686.5 eV) and Li_xPF_y (687.5 eV).

Additionally, the morphological differences between the surfaces of the C-SiO$_x$ and (C-SiO$_x$)@C electrodes during charge/discharge cycles were investigated using SEM and TEM analysis, as observed in **Figure 9**. After 200 cycles, extensive cracking and partial fracture of the C-SiO$_x$ electrode was observed, whereas the surface of the (C-SiO$_x$)@C electrode was stably retained. These results indicate that the benzene-derived multiple-carbon matrix could play an important role in improving the cyclic stability and electrical conductivity of SiO$_x$ to enable its use as a promising anode for LIBs.

3.3 Boosting the performance by Ti doping on SiO$_x$ sites

We previously demonstrated the exceptional improvement of the electrochemical performance of the (C-SiO$_x$)@C composite achieved through the formation of interconnected carbon paths and a homogenous carbon coating. These results confirmed that improvement of the electrical conductivity of the SiO$_x$ composite affected the electrochemical properties. Herein, we prepared a Ti^{3+}-doped and

Figure 9.
SEM images and TEM image of (a)–(c) C-SiO_x electrode and (d)–(f) (C-SiO_x)@C electrode before cycling and after 200 cycles at a current density of 0.5 A g^{-1}.

carbon-incorporated/carbon-coated SiO_x ((C-$Ti_xSi_{1-x}O_y$)@C) composite via the alcoholysis-based reaction using $SiCl_4$, $TiCl_4$, and EG with benzene. The detailed reaction mechanism of the formation of the (C-$Ti_xSi_{1-x}O_y$)@C composite was as follows:

$$xTiCl_4 + (1 - x)\,SiCl_4 + yC_2H_4(OH)_2 + zC_6H_6 \rightarrow Ti_xSi_{1-x}(O_2C_2H_2)_y + zC_6H_6 + \alpha HCl\uparrow \quad (6)$$

$$Ti_xSi_{1-x}(O_2C_2H_2)_y + zC_6H_6\,(> 600^{\circ}C) \rightarrow Ti_xSi_{1-x}(OC)_y + 6zC + yCO\uparrow + (y + 3z)\,H_2\uparrow \quad (7)$$

$$Ti_xSi_{1-x}(OC)_y + 6zC \rightarrow (C - Ti_xSi_{1-x}O_y)@C \text{ composite.} \quad (8)$$

After the precursors were mixed, the solution-state mixture was rapidly converted into mineral-like solids containing Si, Ti, O, and C. In these reactions, $TiCl_4$ should preferentially react with EG to achieve a homogenous dispersion of Ti ions in the (C-$Ti_xSi_{1-x}O_y$)@C composite. Additionally, the use of nonpolar benzene, an additional carbon source, enabled the formation of a homogeneous distribution of the nonpolar $SiCl_4$ and $TiCl_4$ through dispersion forces during the reaction.

As shown in **Figure 10a**, XRD patterns of the (C-$Ti_xSi_{1-x}O_y$)@C composite revealed the presence of typical amorphous phases of SiO_2, which was comparable to the results for SiO_x reported above. The existence of carbon in the sample was verified using Raman spectroscopy. In **Figure 10b**, peaks at 1350 and 1690 cm^{-1}, corresponding to the D band and G band, respectively, were clearly revealed for the (C-$Ti_xSi_{1-x}O_y$)@C composite. To further characterize the composition of the (C-$Ti_xSi_{1-x}O_y$)@C composite, XPS analysis was performed. As shown in **Figure 10c**, two characteristic peaks for Si^{4+} corresponding to the previously synthesized C-SiO_x and (C-SiO_x)@C composite in the Si 2p spectra were observed. However, for the (C-$Ti_xSi_{1-x}O_y$)@C composite, peaks corresponding to Si^{3+} were observed, which resulted from electrons trapped in the Ti^{3+}/Ti^{4+} state or oxygen vacancies. As shown

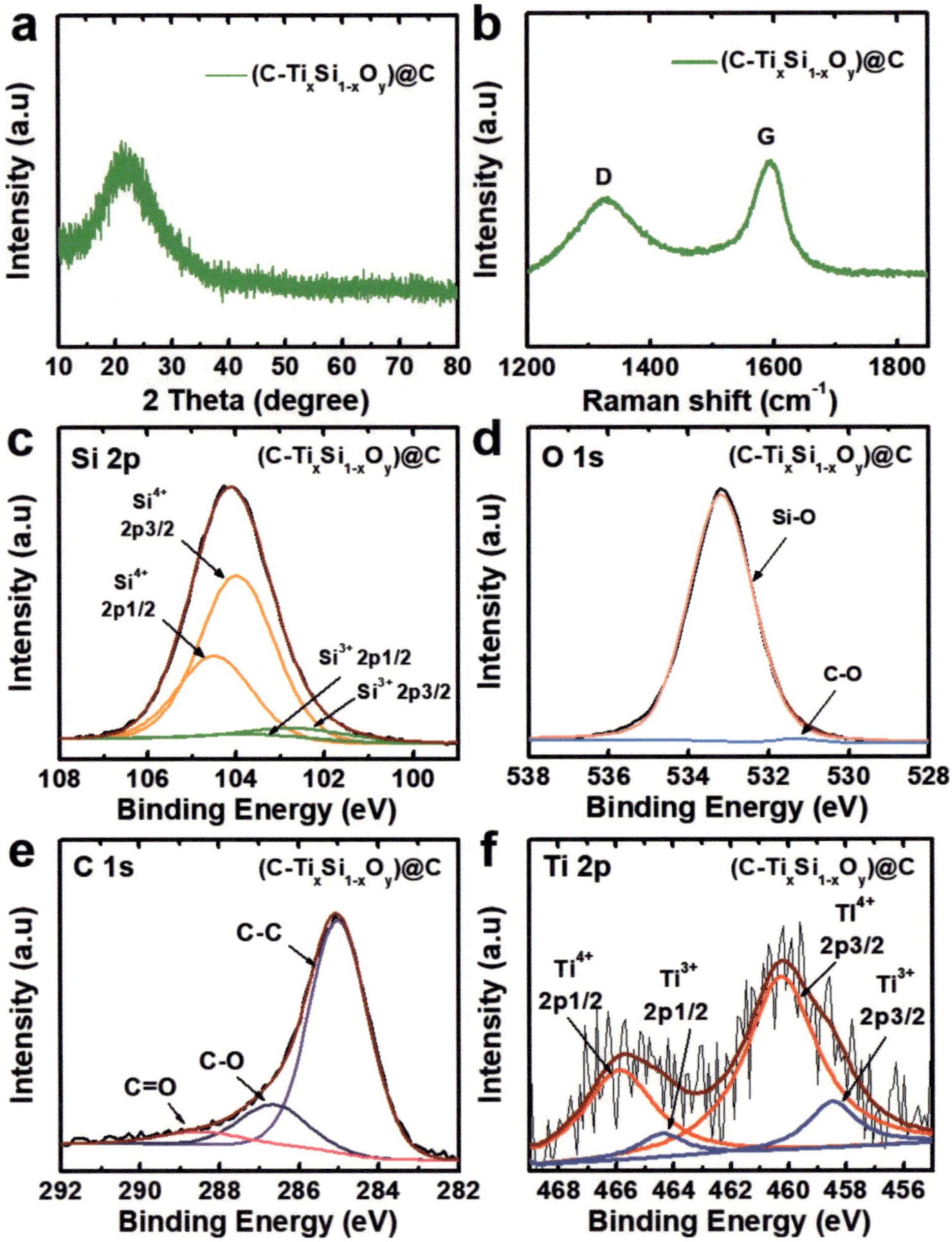

Figure 10.
(a) XRD patterns, (b) Raman spectra, and XPS spectra showing (c) Si 2p, (d) O 1s, (e) C 1s, and (f) Ti 2p of (C-$Ti_xSi_{1-x}O_y$)@C composite.

in **Figure 10d**, Si—O and C—O bonding were detected for the (C-$Ti_xSi_{1-x}O_y$)@C composite. As observed in **Figure 10e**, characteristic peaks were detected in the C 1s spectra of the (C-$Ti_xSi_{1-x}O_y$)@C composite, indicating the existence of the carbon framework in the composites. The existence of Ti in the (C-$Ti_xSi_{1-x}O_y$)@C composite was also clearly confirmed through XPS analysis. In **Figure 10f**, four characteristic peaks appeared in the Ti 2p spectra of the (C-$Ti_xSi_{1-x}O_y$)@C composite. The peaks at ~464.4 and ~458.5 eV correspond to the Ti^{3+} ion of Ti_2O_3, and the peaks at ~460.2 and ~465.8 eV were attributed to the Ti^{4+} ion of TiO_2 [29–31], which indicates the formation of Ti—O bonds in the (C-$Ti_xSi_{1-x}O_y$)@C composite during the heat-treatment process. However, because far fewer Ti ions than Si ions were present in the composite, we suspected that the Ti—O bond peak was covered by the Si—O bond peak in the XPS spectrum. The existence of Ti^{3+} ions implies that the (C-$Ti_xSi_{1-x}O_y$)@C composite could possess improved electrical conductivity compared with that of the C-SiO_x and (C-SiO_x)@C composite.

TEM–EDS analysis was employed to examine the elemental dispersion of the (C-$Ti_xSi_{1-x}O_y$)@C. As displayed in **Figure 11a**, the elements of Si, O, C, and Ti almost overlapped, indicating the homogeneous distribution of Ti in the SiO_x composite. The carbon component also indicated that each particle was encircled by a carbon-rich region with a thickness of ~20 nm. However, it was verified that despite existence of Ti, any peaks were not observed in XRD and Raman of the (C-$Ti_xSi_{1-x}O_y$)@C composite, which indicates that the intensity resulting from Ti is very low and overall phase of this composite is amorphous. Thus, we supposed that Ti was not detected by XRD and Raman, although TEM mapping images and XPS spectrum showed existence of Ti in the composite. The TEM images in **Figure 11b** and **c** confirm the presence of crystalline Si with sizes of ~15 nm and reveal (111) planes with an interplanar spacing of 3.1 Å. Si nanoparticles were distributed in the SiO_x and carbon matrix, contributing to the improved capacity of the (C-$Ti_xSi_{1-x}O_y$)@C composite. Additionally, **Figure 12** presents SEM images of the C-SiO_x, (C-SiO_x)@C, and (C-$Ti_xSi_{1-x}O_y$)@C composite; the morphology and size of the particles were not affected by the addition of benzene and the Ti-based source. The elemental compositions and calculated atomic ratios of oxygen to silicon of the samples are listed in **Table 1**. The O/Si ratio decreased with the addition of benzene and the Ti source, suggesting an increase in the carbon content and electron trapping resulting from the presence of Ti^{3+}/Ti^{4+} ions. These experimental results indicate that the (C-$Ti_xSi_{1-x}O_y$)@C composite was successfully prepared, that the electrical conductivity was enhanced by the presence of Ti^{3+} ions, and that the carbon coating might result in outstanding electrochemical performance of the SiO_x composite.

Figure 13a and **b** present voltage profiles and show the cycling performance, respectively, of the (C-$Ti_xSi_{1-x}O_y$)@C composite at 0.1 A g^{-1} for 100 cycles compared with those of the C-SiO_x and (C-SiO_x)@C composite. The initial discharge capacity of the (C-$Ti_xSi_{1-x}O_y$)@C composite was ~1304 mAh g^{-1}, and the capacity retention was 95.2% after 100 cycles, corresponding to a capacity loss of 0.048% per cycle. To determine the power capability of each sample, the electrochemical properties were measured at various current densities. As observed in **Figure 13c**, at 1 A g^{-1}, up to ~985 mAh g^{-1} of this discharge capacity was retained, which was ~5 times higher than that of C-SiO_x under the same conditions. In addition, as shown in **Figure 13d**, the capacity retentions of the C-SiO_x and (C-SiO_x)@C composites were only 58.9 and 86.8%, respectively, over 600 cycles at 1 A g^{-1} after 5 cycles at 0.1 A g^{-1}. In contrast, up to ~88.9% of the initial discharge capacity of the (C-$Ti_xSi_{1-x}O_y$)@C composite was retained after 600 cycles under the same conditions with a high coulombic efficiency of ~99%.

To further understand the differences in the electrochemical performance of the three electrodes, EIS measurements were performed after 50 cycles at 0.1 A g^{-1} over the frequency range of 1 MHz to 1 mHz with an AC amplitude of 10 mV. As observed in **Figure 14a**, the Nyquist plots of all of the samples consisted of a semicircle at high frequency and a straight line at low frequency. The first intersection of the semicircle at high frequency with the real axis is related to the electrolyte solution resistance (R_{el}), and the diameter of the semicircle is related to the charge-transfer resistance (R_{ct}) resulting from the reaction at the electrode-electrolyte interface. The straight line at low frequency is related to the Warburg impedance (Z_{re}) corresponding to the Li-ion diffusion; the Warburg impedance coefficient

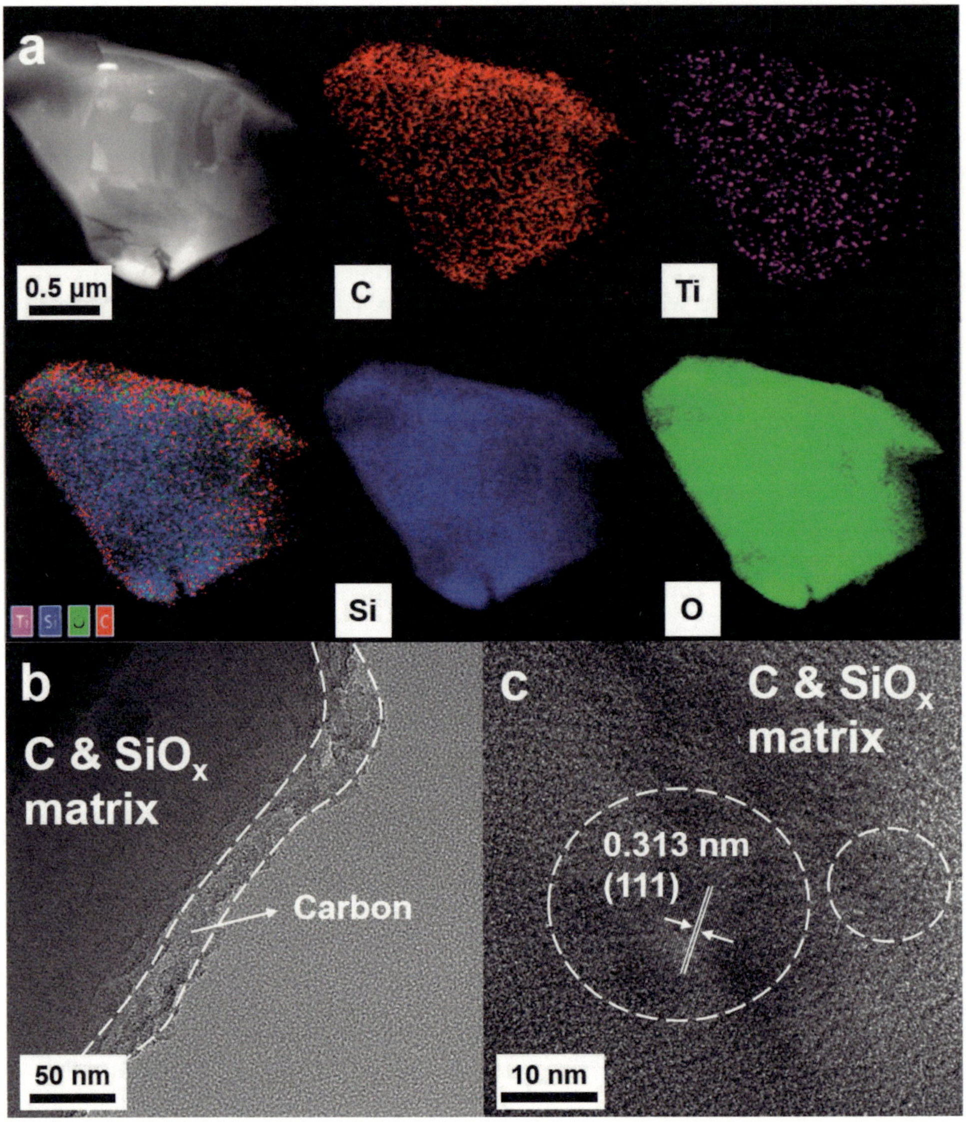

Figure 11.
TEM characterization of (C-$Ti_xSi_{1-x}O_y$)@C composite: (a) TEM elemental mapping images of C, Ti, Si, and O. (b) TEM image of ~20-nm-thick carbon layer on amorphous SiO_2 matrix. (c) TEM image showing lattice fringes of Si nanoparticles. The white dashed circles identify crystalline nano-Si with a planar distance of 3.1 Å at (111).

Figure 12.
SEM images of (a) C-SiO_x, (b) (C-SiO_x)@C composite and (c) (C-$Ti_xSi_{1-x}O_y$)@C composite.

Sample	Element content (wt%)				Atomic ratio
	Si	C	O	Ti	O/Si
C-SiO_x	45.8	7.3	46.9		1.80
(C-SiO_x)@C	46.3	10.4	43.3		1.64
(C-Ti_xSi_1-xO_y)@C	47.2	10.7	41.8	0.25	1.55

Table 1.
Elemental composition and atomic ratio of O/Si of pristine C-SiO_x, (C-SiO_x)@C and (C-Ti_xSi_1-xO_y)@C composite.

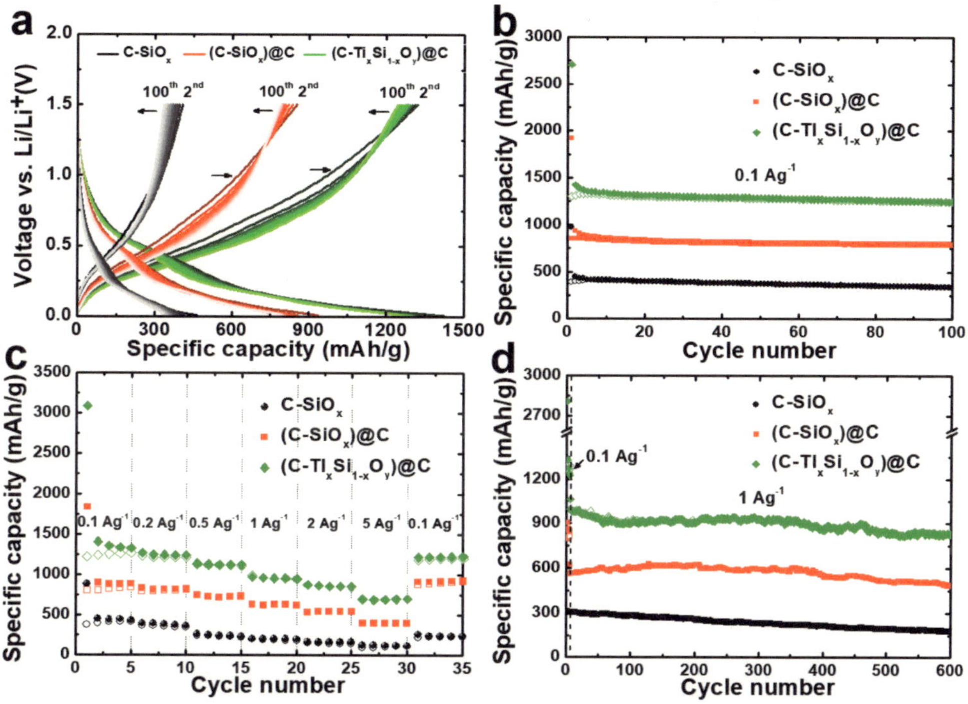

Figure 13.
Electrochemical characteristics of C-SiO_x, (C-SiO_x)@C composite and (C-Ti_xSi_1-xO_y)@C composite. (a) Charge/discharge profiles and (b) cycling performance at current density of 0.1 A g^{-1}. (c) Power capability at various current densities. (d) Cycling performance at a current density of 1 A g^{-1}.

(σ_w) can be estimated from the relations between Z_{re} and $\omega^{-1/2}$ at low frequency using Eq. (9), where ω is the angular frequency [32–34].

$$Z_{re} = R_{ct} + R_{el} + \sigma_w \omega^{-1/2} \quad (9)$$

The Li-ion diffusion coefficient (D^{Li}) can be calculated using Eq. (10), where R, T, A, n, F, and C refer to the gas constant, temperature, surface area, number of electrons per molecule participating in the redox reaction, Faraday constant, and maximum ion concentration (7.69×10^{-3} mol cm^{-3}), respectively.

$$D_{Li} = R^2T^2/2A^2n^4F^4C^2\sigma^2 \quad (10)$$

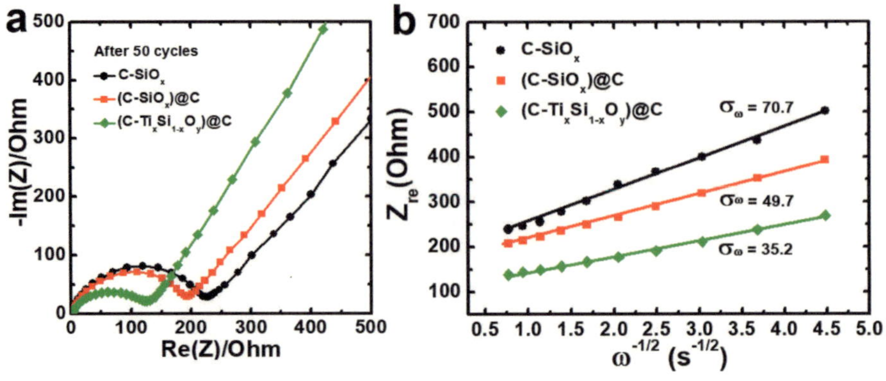

Figure 14.
(a) Nyquist plots of EIS results and (b) Z_{re}–$\omega^{-1/2}$ plots in the low-frequency range for C-SiO$_x$, (C-SiO$_x$)@C composite, and (C-Ti$_x$Si$_1$–xO$_y$)@C composite after 50 cycles at a current density of 0.1 A g^{-1} over the frequency range of 1 MHz to 1 mHz with an AC amplitude of 10 mV.

	C-SiO$_x$	(C-SiO$_x$)@C	(C-Ti$_x$Si$_1$−xO$_y$)@C
R_{el}/Ω	1.023	1.045	1.124
R_{ct}/Ω	230.7	204.9	134.9
$\sigma_w/\Omega\ s^{-1/2}$	70.7	49.7	35.2
$D_{Li}/cm^2\ s^{-1}$	5.06×10^{-14}	1.02×10^{-13}	2.04×10^{-13}

Table 2.
R_{el}, R_{ct}, σ_w, and Li-ion diffusion coefficients of C-SiO$_x$, (C-SiO$_x$)@C composite, and (C-Ti$_x$Si$_1$–xO$_y$)@C composite.

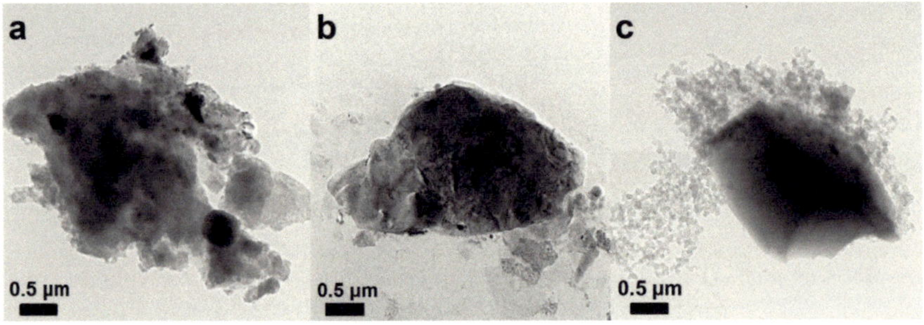

Figure 15.
TEM images of (a) C-SiO$_x$, (b) (C-SiO$_x$)@C composite, and (c) (C-Ti$_x$Si$_1$–xO$_y$)@C composite electrodes after 300 cycles at a current density of 1 A g^{-1}.

As shown in **Table 2**, the calculated D_{Li} of the (C-Ti$_x$Si$_1$–xO$_y$)@C composite was 2.04×10^{-13} cm^2 s^{-1}, which indicates that the Li-ion diffusion was ~4 times faster than that for C-SiO$_x$ (5.06×10^{-14} cm^2 s^{-1}). Furthermore, R_{ct} of the C-SiO$_x$, (C-SiO$_x$)@C, and (C-Ti$_x$Si$_1$–xO$_y$)@C electrodes were determined to be 230.7, 204.9, and 134.9 Ω, respectively. Thus, it can be concluded that the multiple structural modifications resulting from the Ti doping and carbon coating led to increased D_{Li} and reduced R_{ct} of the SiO$_x$ composite, indicating that the (C-Ti$_x$Si$_1$–xO$_y$)@C composite could exhibit highly enhanced electrochemical performance as a promising anode for LIBs.

The stable cycle life of the (C-Ti$_x$Si$_1$–xO$_y$)@C composite resulted from its high structural stability because of the multiple-carbon matrix derived from EG and benzene. As shown in **Figure 15a–c**, whereas severe particle degradation was detected for C-SiO_x after 300 cycles at high current density, the morphologies of the (C-SiO_x)@C and (C-$Ti_xSi_{1-x}O_y$)@C composites were retained under the same conditions. Thus, the (C-SiO_x)@C and (C-$Ti_xSi_{1-x}O_y$)@C composites exhibited improved structural stability compared with C-SiO_x. The fracture of particles during cycling leads to the loss of electrical contact and increase of the charge-transfer resistance, resulting in gradual capacity fade during cycling. Therefore, these results indicate that the (C-SiO_x)@C and (C-$Ti_xSi_{1-x}O_y$)@C composites would exhibit improved cyclability compared with C-SiO_x.

4. Conclusion

In summary, we successfully fabricated SiO_x active materials using a simple and cost-effective one-pot synthesis method via an alcoholysis-based reaction. We also demonstrated the exceptional improvement of the electrochemical performance of (C-$Ti_xSi_{1-x}O_y$)@C compared with that of SiO_x achieved by structural modifications using Ti doping and a multiple-carbon matrix. The (C-$Ti_xSi_{1-x}O_y$)@C composite consisted of uniformly dispersed Ti ions in an amorphous carbon and SiO_x matrix, which was homogenously encapsulated by ~20-nm-thick carbon, enabling the achievement of high power capability and outstanding cyclability. At 1 A g^{-1}, the (C-$Ti_xSi_{1-x}O_y$)@C electrode retained a discharge capacity of up to ~995 mAh g^{-1}, which was ~3 times higher than that retained by C-SiO_x under the same conditions. Furthermore, the structural modifications also provided an effective buffer that prevented the severe structural degradation caused by the large volumetric expansion during the charge/discharge cycles. As a result, the (C-$Ti_xSi_{1-x}O_y$)@C composite exhibited superior cycle life stability. The C-SiO_x and (C-SiO_x)@C composite retained up to ~58.9 and ~86.8% of their initial capacities after 600 cycles at 1 A g^{-1}, respectively, whereas the (C-$Ti_xSi_{1-x}O_y$)@C composite delivered a capacity retention of ~88.9%.

Acknowledgements

This work is supported by the National Research Foundation of Korea (NRF) grant funded by the Ministry of Science, ICT and Future Planning (NRF-2016R1A2B4014521) and (NRF-2015M3D1A1069713).

Conflict of interest

The authors declare no competing financial interest.

Notes/thanks/other declarations

We obtained permission for the figures and tables used in this paper from Journal of The Electrochemical Society and Journal of Power Sources.

Author details

Hyeon-Woo Yang and Sun-Jae Kim*
Department of Nanotechnology and Advanced Materials Engineering, Sejong University, Seoul, Republic of Korea

*Address all correspondence to: sjkim1@sejong.ac.kr

References

[1] Liu N, Lu Z, Zhao J, McDowell MT, Lee H-W, Zhao W, et al. A pomegranate-inspired nanoscale design for large-volume-change lithium battery anodes. Nature Nanotechnology. 2014;**9**:187-192. DOI: 10.1038/NNANO.2014.6

[2] Obrovac M, Chevrier V. Alloy negative electrodes for Li-ion batteries. Chemical Reviews. 2014;**114**: 11444-11502. DOI: 10.1021/cr500207g

[3] Choi S, Kwon T-W, Coskun A, Choi JW. Highly elastic binders integrating polyrotaxanes for silicon microparticle anodes in lithium ion batteries. Science. 2017;**357**:279-283. DOI: 10.1126/science.aal4373

[4] Casimir A, Zhang H, Ogoke O, Amine JC, Lu J, Wu G. Silicon-based anodes for lithium-ion batteries: Effectiveness of materials synthesis and electrode preparation. Nano Energy. 2016;**27**:359-376. DOI: 10.1016/j.nanoen.2016.07.023

[5] Xing Y, Shen T, Guo T, Wang X, Xia X, Gu C, et al. A novel durable double-conductive core-shell structure applying to the synthesis of silicon anode for lithium ion batteries. Journal of Power Sources. 2018;**384**:207-213. DOI: 10.1016/j.jpowsour.2018.02.051

[6] Shang H, Zuo Z, Yu L, Wang F, He F, Li Y. Low-temperature growth of all-carbon graphdiyne on a silicon anode for high-performance lithium-ion batteries. Advanced Materials. 2018;**30**:1801459. DOI: 10.1002/adma.201801459

[7] Li Z, He Q, He L, Hu P, Li W, Yan H, et al. Self-sacrificed synthesis of carbon-coated SiO_x nanowires for high capacity lithium ion battery anodes. Journal of Materials Chemistry A. 2017;**5**: 4183-4189. DOI: 10.1039/x0xx00000x

[8] Parimalam BS, Mac Intosh AD, Kadam R, Lucht BL. Decomposition reactions of anode solid electrolyte interphase (SEI) components with $LiPF_6$. The Journal of Physical Chemistry C. 2017;**121**:22733-22738. DOI: 10.1021/acs.jpcc.7b08433

[9] Haruta M, Okubo T, Masuo Y, Yoshida S, Tomita A, Takenaka T, et al. Temperature effects on SEI formation and cyclability of Si nanoflake powder anode in the presence of SEI-forming additives. Electrochimica Acta. 2017;**224**:186-193. DOI: 10.1016/j.electacta.2016.12.071

[10] Elia GA, Hassoun J. A SiO_x-based anode in a high-voltage lithium-ion battery. ChemElectroChem. 2017;**4**:2164-2168. DOI: 10.1002/celc.201700316

[11] Park E, Park MS, Lee J, Kim KJ, Jeong G, Kim JH, et al. A highly resilient mesoporous SiO_x lithium storage material engineered by oil–water templating. ChemSusChem. 2015;**8**: 688-694. DOI: 10.1002/cssc.201402907

[12] Sun L, Su T, Xu L, Liu M, Du H-B. Two-dimensional ultra-thin SiO_x ($0 < x < 2$) nanosheets with long-term cycling stability as lithium ion battery anodes. Chemical Communications. 2016;**52**:4341-4344. DOI: 10.1039/c6cc00723f

[13] Guo C, Wang D, Liu T, Zhu J, Lang X. A three dimensional SiO_x/C@rGO nanocomposite as a high energy anode material for lithium-ion batteries. Journal of Materials Chemistry A. 2014;**2**:3521-3527. DOI: 10.1039/c3ta13746e

[14] Zhang J, Zhang C, Liu Z, Zheng J, Zuo Y, Xue C, et al. High-performance ball-milled SiO_x anodes for lithium ion batteries. Journal of Power Sources. 2017;**339**:86-92. DOI: 10.1016/j.jpowsour.2016.11.044

[15] Shi L, Wang W, Wang A, Yuan K, Jin Z, Yang Y. Scalable synthesis of core-shell structured SiO_x/nitrogen-doped carbon composite as a high-performance anode material for lithium-ion batteries. Journal of Power Sources. 2016;**318**: 184-191. DOI: 10.1016/j.jpowsour. 2016.03.111

[16] Xu Q, Sun JK, Yin YX, Guo YG. Facile synthesis of blocky SiO_x/C with graphite-like structure for high-performance lithium-ion battery anodes. Advanced Functional Materials. 2018;**28**:1705235. DOI: 10.1002/ adfm.201705235

[17] Park E, Yoo H, Lee J, Park M-S, Kim Y-J, Kim H. Dual-size silicon nanocrystal-embedded SiO_x nanocomposite as a high-capacity lithium storage material. ACS Nano. 2015;**9**:7690-7696. DOI: 10.1021/ acsnano.5b03166

[18] Liu Q, Cui Z, Zou R, Zhang J, Xu K, Hu J. Surface coating constraint induced anisotropic swelling of silicon in Si–Void@SiO_x nanowire anode for lithium-ion batteries. Small. 2017;**13**:1603754. DOI: 10.1002/smll.201603754

[19] Han J, Chen G, Yan T, Liu H, Shi L, An Z, et al. Creating graphene-like carbon layers on SiO anodes via a layer-by-layer strategy for lithium-ion battery. Chemical Engineering Journal. 2018;**347**:273-279. DOI: 10.1016/j. cej.2018.04.100

[20] Dou F, Shi L, Song P, Chen G, An J, Liu H, et al. Design of orderly carbon coatings for SiO anodes promoted by TiO_2 toward high performance lithium-ion battery. Chemical Engineering Journal. 2018;**338**:488-495. DOI: 10.1016/j.cej.2018.01.048

[21] Lee K-M, Lee Y-S, Kim Y-W, Sun Y-K, Lee S-M. Electrochemical characterization of Ti–Si and Ti–Si–Al alloy anodes for Li-ion batteries produced by mechanical ball milling. Journal of Alloys and Compounds. 2009;**472**:461-465. DOI: 10.1016/j. jallcom.2008.04.102

[22] Wang Y, He Y, Xiao R, Li H, Aifantis K, Huang X. Investigation of crack patterns and cyclic performance of Ti–Si nanocomposite thin film anodes for lithium ion batteries. Journal of Power Sources. 2012;**202**:236-245. DOI: 10.1016/j.jpowsour.2011.11.027

[23] Ren Y, Li J, Yu J. Enhanced electrochemical performance of TiO_2 by Ti^{3+} doping using a facile solvothermal method as anode materials for lithium-ion batteries. Electrochimica Acta. 2014;**138**:41-47. DOI: 10.1016/j. electacta.2014.06.068

[24] Seok D-I, Wu M, Shim KB, Kang Y, Jung H-K. High-rate performance of Ti^{3+} self-doped TiO_2 prepared by imidazole reduction for Li-ion batteries. Nanotechnology. 2016;**27**:435401. DOI: 10.1088/0957-4484/27/43/435401

[25] Chen J, Song W, Hou H, Zhang Y, Jing M, Jia X, et al. Ti^{3+} self-doped dark rutile TiO_2 ultrafine nanorods with durable high-rate capability for lithium-ion batteries. Advanced Functional Materials. 2015;**25**:6793-6801. DOI: 10.1002/adfm.201502978

[26] Miyachi M, Yamamoto H, Kawai H, Ohta T, Shirakata M. Analysis of SiO anodes for lithium-ion batteries. Journal of the Electrochemical Society. 2005;**152**:A2089-A2091. DOI: 10.1149/1.2013210

[27] Philippe B, Dedryvère RM, Allouche J, Lindgren F, Gorgoi M, Rensmo HK, et al. Nanosilicon electrodes for lithium-ion batteries: Interfacial mechanisms studied by hard and soft x-ray photoelectron spectroscopy. Chemistry of Materials. 2012;**24**:1107, 1115. DOI: 10.1021/cm2034195

[28] Verma P, Maire P, Novák P. A review of the features and analyses

of the solid electrolyte interphase in Li-ion batteries. Electrochimica Acta. 2010;**55**:6332-6341. DOI: 10.1016/j.electacta.2010.05.072

[29] Hashimoto S, Tanaka A. Alteration of Ti 2p XPS spectrum for titanium oxide by low-energy Ar ion bombardment. Surface and Interface Analysis. 2002;**34**:262-265. DOI: 10.1002/sia.1296

[30] Chrcanovic BR, Pedrosa AR, Martins MD. Chemical and topographic analysis of treated surfaces of five different commercial dental titanium implants. Materials Research. 2012;**15**:372-382. DOI: 10.1590/S1516-14392012005000035

[31] Park AR, Son D-Y, Kim JS, Lee JY, Park N-G, Park J, et al. Si/Ti_2O_3/reduced graphene oxide nanocomposite anodes for lithium-ion batteries with highly enhanced cyclic stability. ACS Applied Materials & Interfaces. 2015;7: 18483-18490. DOI: 10.1021/acsami.5b04652

[32] Jiang Z, Pei B, Manthiram A. Randomly stacked holey graphene anodes for lithium ion batteries with enhanced electrochemical performance. Journal of Materials Chemistry A. 2013;**1**:7775-7781. DOI: 10.1039/c3ta10457e

[33] Li S, Cao X, Schmidt CN, Xu Q, Uchaker E, Pei Y, et al. $TiNb_2O_7$/graphene composites as high-rate anode materials for lithium/sodium ion batteries. Journal of Materials Chemistry A. 2016;**4**:4242-4251. DOI: 10.1039/c5ta10510b

[34] Tan Y, Wong KW, Ng KM. Novel silicon doped tin oxide–carbon microspheres as anode material for lithium ion batteries: The multiple effects exerted by doped Si. Small. 2017;**13**:1702614. DOI: 10.1002/smll.201702614

Chapter 6

Progress on Free-Standing Graphene Hybrid: Advantages and Future Scenario

Karthick Ramalingam and Fuming Chen

Abstract

Free-standing graphene (FSG) paper like electrodes has paid attention to the energy storage device application in the past decade. It befits to fabricate flexible devices due to its remarkable mechanical strength and offers high electrical conductivity. In this chapter, we explore the advantages and future prospects of FSG fresh candidate in rechargeable batteries. Herein, we summarized the synthetic strategies used for FSG fabrication and its properties, followed by its application in rechargeable batteries. Extensively, this chapter deals with fabrication of FSG hybrid composite papers for battery applications to understand the overall device performance. Specifically, we discuss the benefits of FSG electrodes over conventional electrode material and its fabrication in battery system. Ultimately, we conclude with the significance of FSG paper in battery application and forthcoming advantage for recycling purposes.

Keywords: free-standing graphene paper, electroactive, Li-ion battery, Na-ion battery, Li-S battery, Li-air battery

1. Introduction

Pertaining to the day-to-day energy usage increases, various technologies were addressed to satisfy the current energy demand. Based on this circumstance, the electronic devices for energy conversion (solar cells and fuel cells) and energy storage (batteries and supercapacitors) were extensively studied throughout the world [1]. Basically, the performance of these devices depends on the materials' design with different nanostructures and material interfaces. In particular, advanced materials including carbon nanomaterials, viz., carbon black, carbon nanotubes, carbon nanofibers, graphene, and so on, play a vital role in an attempt to lead the breakthrough and challenges from laboratory scale to technology ideas [2].

Among them, graphene, since its discovery, has been stirring enthusiasm among the scientific community owing to its attractive properties. Properties such as high electrocatalytic activity, good conductivity with immense surface area, and low costs make it an ideal candidate to implement in electrochemical application. Subsequently, graphene has been utilized as a promising candidate in energy storage applications such as battery and supercapacitors (SCs) [3, 4]. Due to its high electrical conductivity, charge carrier mobility, and transparency, it has been potentially used as an electrode for electrochemical energy device application [5, 6]. Processing of graphene electrodes differs according to their application by fabrication techniques and synthetic strategies. As graphene is an electrode focusing on rechargeable battery application, the device performance is based on the presence of electroactive sites in graphene sheets [7, 8]. Therefore, graphene sheets composited with suitable electroactive materials like metal chalcogenides, metal oxides/hydroxides, metal nanostructures, and even the heteroatom-doped graphene provide better activity for rechargeable batteries [9–11]. Conventionally, the electrode materials were deposited on metal foils by doctor-blade technique, drop-casting, spray-coating, or spin coating to construct the batteries. This electrode material was mixed with foreign materials (binders and conducting agent) to make into ink, paste, colloidal dispersion, etc., for deposition purposes. In the case of self-supported graphene foams or FSGs, the foreign materials are avoided, and on the whole, they act as electrodes directly [12]. This chapter outlines few reported literature on FSG performance for rechargeable battery applications. Moreover, we summarized the synthetic strategies and fabrication of free-standing graphene/hybrid functional materials for particular device application.

2. Graphene: properties and nomenclature

Graphene is a 2D one atom thin sheet that consists of hexagonal sp^2 carbon, which is densely packed into honey-comb lattice and large benzene-like aromatic hydrocarbon. It is considered as fundamental basis for all carbon allotropes, and their conceptual depiction are shown in **Figure 1**. It represents that 2D graphene sheet can be enclosed into 0D like fullerene structure and rolled up into 1D-like carbon nanotube structure, and 10 layers of graphene can be stacked up into 3D graphitic-like structure. Hence, it is considered as "mother of carbon allotropes" [13]. The fabrication of graphene film by different synthetic routes was adapted accordingly to its required

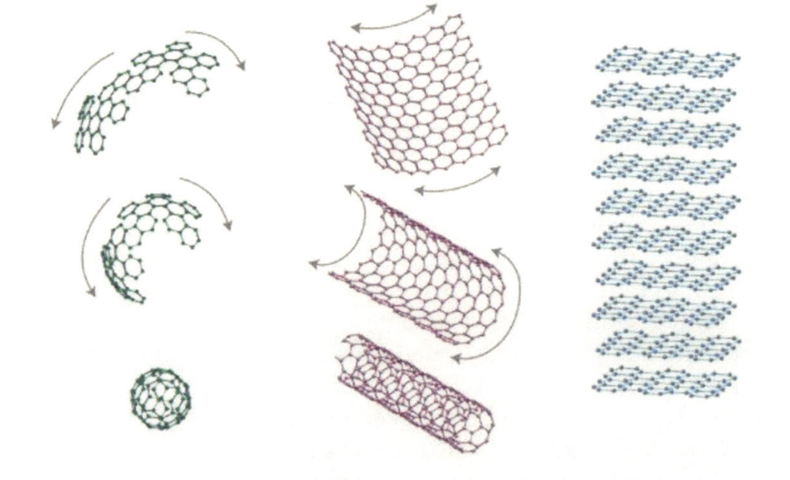

Figure 1.
Carbon allotropes in different forms: 0D Bucky ball, 1D nanotubes, 2D sheets, and 3D graphite form (without permission from Ref. [13]).

properties for many applications. Current technologies addressed to synthesize graphene via several routes are as follows: mechanical exfoliation (liquid exfoliation and scotch tape method), epitaxial growth (chemical vapor deposition (CVD) and from organic molecules method), unzipping CNT (chemical and electrochemical methods), and wet chemical process (oxidation of graphite) [14].

2.1 Graphene: properties

Graphene possesses exclusive chemical, physical, mechanical, and thermal properties, which focuses on the field of electrochemical applications as an electrode material to enhance the stability and durability of the devices. Graphene application in any devices is adopted according to its properties as shown in **Figure 2**. Prominently, the conductivity of anode and cathode electrodes plays a vital role in batteries, which collect or disperse the electrons that tune up the performance to device. The conjugated sp^2 carbon networks of 2D graphene sheet exhibit high conductivity around 10^4–10^6 S/cm than any other carbon materials depending on the number of layers [15, 16]. Additionally, the electrode surface area is an essential part for batteries, which has high theoretical surface area of graphene, and is reported to be ~2600 m^2/g [17]. For suspended graphene sheets below 10 nm thickness, the spring constants were observed between 1 and 5 N/m, and pristine graphene exhibits Young's modulus of 1.05 TPa and intrinsic strength of 110 GPa, which has high mechanical property [18, 19]. The electrochemical property is a perspective for energy storage and generation technologies. The rate of heterogeneous electron transfer occurs on graphene materials; in the meantime, the rate of

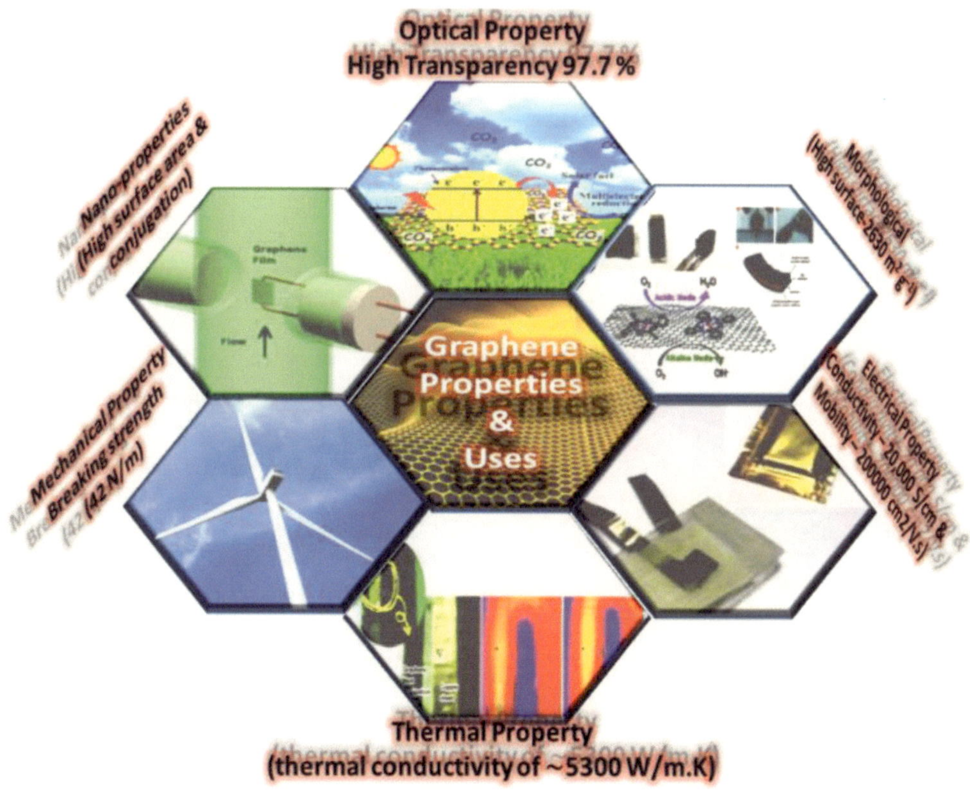

Figure 2.
Properties of graphene and its appropriate application.

reaction varies selectively at edges and basal plane according to their electroactive sites by adding impurities or doping. Graphene-based materials were potentially applied in electrochemical devices due to their inherent electrochemical activity nature [20]. These amazing properties of graphene such as electrical, mechanical, and electrochemical were attracted for rechargeable batteries.

2.2 Graphene: nomenclature

It is well known that graphene can be synthesized by several routes and named according to the recovered final product. Graphene research has elevated gradually in the past 5 years for its tremendous properties, but the scientific community ends up with the confusion in naming the material. Even though researchers have synthesized up to 100 layers of carbon sheets, they were naming them as graphene. This provides different changes in properties compared with the single-layer graphene sheet for their practical applications [21]. Hence, carbon journal community raised a nomenclature for graphene family, which is shown in (**Table 1**).

The descriptive term is an essential thing for researchers in the area of graphene material because the properties will change accordingly with recovered product with different synthetic strategies. For example, the graphene-based transparent conducting film adopted by the CVD method obtained 600 ohms/sq. at 96.5% transmittance at 550 nm, whereas solution processed graphene increases above 10 K ohms at the same transmittance [22–24]. Even the electrochemical behavior fluctuates according to the synthetic strategies; for instance, the presence of oxygen functional groups in graphene oxide (GO) shows an excellent electrochemical behavior rather than the pristine graphene [25]. Hence, the electrochemical device

Materials	Description
Graphene	Two-dimensional sheet with one atom thickness
Turbostratic graphene	Arrangement of graphene sheets in rotational fault structure
Bi-,tri-, or multilayer graphene	Stacking of graphene sheets (2 - bi, 3 - tri, & 4 - 10 – multi) in AB, ABA, or rotational order
Few layer graphene	Subset of multilayer graphene
Graphite nanosheets, nanoflakes, and nanoplates	Lateral/thickness of graphene sheets <100 nm.
Exfoliated graphite	Exfoliation of bulk graphite
Graphene nanoribbon	Length dimension in micron and width in the range of nanometer
Graphene quantum dots	Lateral dimension less than 10 nm with photoluminescence property
Graphene oxide	Graphene sheets that contain functional groups (epoxy, hydroxyl, and carboxyl)
Graphite oxide	Exfoliation of bulk graphite by strong oxidation process
Reduced graphene oxide	Reduction or restoration of sp^2 carbon of graphene oxide
Graphenization	Growth of graphene by small molecules (bottom-up approach)
Free-standing graphene, graphene foam, hydrogel, and aerogel	Graphene sheets arranged in 3D forms

Table 1.
Nomenclature of graphene based on the structure.

applications based on graphene electrodes depend on the architecture and hybrid composites to improve the active sites. Recently, 3D architecture like graphene materials such as foams, hydrogel, aerogel, and free-standing was utilized in electrochemistry-oriented topics.

3. Free-standing graphene: synthesis and its properties

For designing and fabricating large scale macroscopic or microscopic architecture like materials, the choice of precursor signifies the synthetic strategies. Graphene sheets synthesized by wet chemical process commenced for several applications due to the presence of functional groups. As discussed in the previous section, the methods utilized for the preparation of graphene sheets conclude their suitable application based on their properties. Noteworthy, there is a challenge for high dispersion of graphene either in aqueous or in organic solvents. It has been achieved by dispersing agent introduced into hydrophobic graphene sheets for good dispersion, whereas it submerges the graphene properties [26]. In the view of fact, large scale solution processable GO has several advantages such as cost effective, eco-friendly solvent and facile to introduce any foreign material due to the presence of functional groups [27, 28]. The copious amount of functional groups attached to the graphene surface contains hydroxyl and epoxy groups at basal planes and carboxyl groups at edges. This leads to affinity with water molecules, which provides a higher dispersion and further it assists with other inorganic or organic molecules for facile composite preparation. In the choice of precursor for free-standing material preparation, GO dominates as a building block due to its features of large scale solution processable with high colloidal dispersion. The resultant macroscopic FSG holds as an excellent mechanical, electrical, and light-weight material. Further, the 3D architecture of FSG enhances the surface area, porous nature, and structural active sites by merging with other functional host materials such as semiconducting material, metal nanoparticles, and polymers. The synergy of graphene sheets and functional host materials in the 3D macroscopic architecture attracted wide variety of applications due to the tuning of their properties.

In 1998, Smalley prepared CNT buckypaper by vacuum filtration, in prior it is well dispersed in Triton X-100 surfactant to break up the pi-pi interaction between the bundled ropes of CNT [29]. Further, CNT buckypapers were prepared by domino pushing technique, and they are strong, robust, and flexible. The obtained paper exhibits 26 micron thickness; the electrical conductivity was found to be 2.0×10^4 S/m and thermal conductivity shows 153 W/mK [30]. These papers were directly applied for supercapacitor application. Thus, the carbon paper–like materials were potentially applied in a variety of applications due to their light-weight, highly flexible, robust, and eco-friendly nature. On the basis of cost, the CNT papers lag behind for the practical applications, and they have been replaced by graphene sheets. Similar to CNT buckypaper, GO paper was fabricated by flow-assisted vacuum filtration or evaporation techniques. **Figure 3a** and **b** shows the photograph of flexible GO paper and mechanical properties comparison chart of GO paper, buckypapers, vermiculite paper-like material, and graphite foil, respectively. Young's modulus is as high as in GO papers with 42 GPa for vacuum-assisted technique, and similar tensile strength but lowest Young's modulus (12.7 GPa) was obtained for evaporation-induced self-assembly technique [31, 34]. Thus, the high mechanical properties of GO paper can be used in several applications such as supercapacitors and other flexible substrates [35]. Moreover, the mechanical properties of GO papers depend on the alignment of GO sheets by any chemical modification between the layers and at the edges. The modifications are made either by crosslinking or grafting between

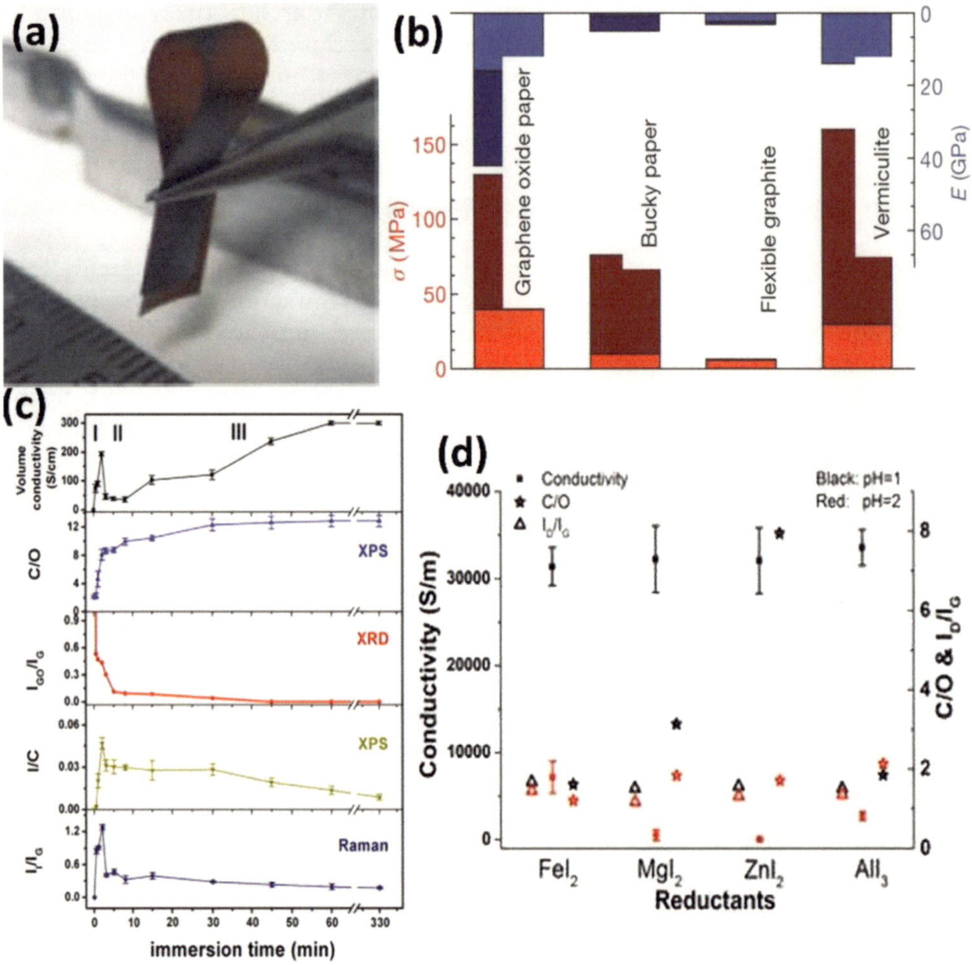

Figure 3.
(a) Photograph of flexible graphene oxide paper, (b) comparison chart of mechanical properties of GO paper with other flexible paper materials, (c) effect of FSG electrical conductivity changes w.r.t its properties upon HI treatment in different scale of time, and (d) electrical conductivity versus the Raman and XPS data of GO paper reduced by different metal halides (without permission from Refs. [31–33]).

the two sheets as GO has several functional groups that covalently attached to other molecules [36, 37]. The intercalation, functionalization, and interaction between the GO sheets provide high mechanical stiffness for paper-like material. Moreover, the atmospheric humidity affects the mechanical property of the GO paper, increase in the relative humidity to 100%, the GO colloidal solution absorbs water from moisture and it bulges to 70% which decreases the tensile strength [34]. The functionalization on graphene surface also affects the mechanical properties depending on the functional moieties as well as the bonding nature [38–40]. The electrical properties of GO papers depend on the synthetic methods as several changes were observed in structures and reduction ratios of C/O. Upon exposing to the hydrazine vapor, the conductivity of GO papers increased by four order of magnitude from 8.5×10^{-4} to 170 S/cm. Further enhancement in conductivities of GO paper was developed by treating the paper with mixture of argon/hydrogen/hydrazine vapors [41]. The removal of the oxygen group is the main factor to restore the sp^2 carbon network by chemical or thermal treatment. The chemical reductive treatment efficiently removes the oxygen moieties from the GO paper, whereas the thermal treatment shows high restoration of sp^2 carbon network but less removal of oxygen functional groups.

Recently, a rapid reduction treatment was proposed by immersing the GO papers in hydrohalic acids, viz., HI and HBr, which shows a remarkable electrical conductivity around 298 and 3220 S/cm, respectively [32, 42]. Based on the facile chemical treatment, the electrical conductivity of FSG improvement was shown by treating the GO papers in metal halides like MgI_2, AlI_3, ZnI_2, and FeI_2 that exhibit 550 S/cm [33].

Owing to these attractive mechanical and electrical properties of FSG material, it played vital role in flexible device technologies based on electrochemical energy storage and generation, actuators, sensors, and catalysts. Based on the attractive graphene properties and its nomenclature, the graphene oxide has fascinating properties which has layered structure similar to graphene that containing oxygen functional groups such as carboxyl, hydroxyl and epoxy. These functional groups were highly dispersed in DI water; hence, it is well aligned over vacuum filtration process. The GO paper is peeled off after vacuum drying and subjected to reducing treatment, as synthesized FSG material is directly utilized as current collector in place of Al, Cu, Ni foam, etc., for energy storage applications.

4. Free-standing graphene electrodes for batteries

Battery is an electrochemical energy storage device that is cost-effective and eco-friendly and with cyclic durability, excellent overall performance, and long-term stability. In this decade, lithium ion battery (LIB) is successfully commercialized worldwide for portable electronic devices, and it has approximately 200 kWh scale for transportation and stationary storage [43]. On comparison with other secondary-based batteries such as sodium sulfur, redox flow, Ni-Cd, etc., Li ion cells have gathered the most commercial interest because they provide high energy and power densities, respectively. In contrast, other secondary batteries are under development stage for consideration in commercial package over LIB due to its major drawback as follows: large scale storage, cost of materials, toxicity, cyclic performance, or stability issues. However, the better system in secondary batteries credited for LIB because the redox potential of −3.04 V vs. SHE (standard hydrogen electrode) for Li/Li+ which has high electropositive in periodic table and light weight material with small ionic radius. Henceforth, the charge-discharge rates enhance and power densities vary in the ranges of 500–2000 W/kg [44]. In commercialized LIBs, the existing negative electrode is a graphite-layered structure material coupled with the host material and $LiCoO_2$ has positive electrodes. Similar to LIBs, the other systems were also focused since it lags behind to reach the theoretical specific capacity (400 Wh/kg) that requires for electric vehicles for long term usage. Hence, other kinds of secondary batteries have been discovered such as Li-sulfur, sodium-ion battery (SIB), sodium-sulfur, Li-air, Zn-air, and flow batteries.

4.1 Li-ion battery

Conventionally, LIBs are made up of graphite anode and $LiCoO_2$ layered material as cathode sandwiched between $LiPF_6$ (1.0 mol/L) as an organic electrolyte dissolved in ethylene carbonate (EC) and dimethyl carbonate (DMC) in 1:1 volume ratio [45]. While LIB is charging, deintercalation happens at cathode, where the Li ions are removed from the layered $LiCoO_2$ by releasing electrons to cathode. The released Li ions are transported to anode with the help of the electrolyte system and finally intercalated into graphite by gaining electrons. The same process is reversed during the discharging process.

Designing of anode materials for LIBs has focused much attention on retaining large reversible specific capacity. Beyond the graphite anode, few metal oxides and metal alloys were developed as anode material, and the lithiation and delithiation processes were investigated. Specifically, FSG paper outpaces the other candidates such as carbon nanotube (CNT) paper or graphite foil due to their tremendous properties as discussed earlier. Importantly, the electrical and mechanical properties of FSG are potentially applied for flexible device application. However, the FSG electrode itself does not provide higher capacity (approximately 100 mAh/g), which is not applicable as anode in LIB; instead, it has good cycling stability. Therefore, the host material that has high electrochemical active sites is incorporated into FSG for improvement of capacity in the device. This extends the large volume expansion in FSG electrodes for an efficient Li ions intercalation. One of the advantages of this FSG hybrid electrode is that it excludes the nonconducting polymer binders as additives. Conventional electrode-based materials were obtained as powders and coated on the metal foils in the form of ink using additives like polymer binders and conducting additive, whereas the FSG hybrid electrode plays dual role as a current collector and conductive additive.

In 2005, LIBs were fabricated with free-standing electrode based on CNTs prepared by vacuum filtration method [46]. Significantly, the free-standing electrode fabrication is a facile route in comparison with the conventional electrode since the mixture of active material, polymer binder, and conductive additive in solvent coated on metal foils. The CNT free-standing electrode provides reversible discharge capacity of 200 mAh/g at 0.08 mA/cm^2. Further, the specific capacity was enhanced by the CVD grown free-standing CNT that delivers 572 mAh/g at 0.2 mA/cm^2 [47]. This is a quite interesting result obtained for free-standing electrodes rather than the conventional electrodes. Meanwhile, the usage of high-cost material CNTs as free-standing electrodes lags behind manufacturing process. From this point of view, inexpensive material graphene prepared by chemical methods provides large scale production as dispersion in many solvents. This dispersion is readily subjected to vacuum filtration to prepare FSG paper with desired thickness. Usually, the discharge capacity of 298 mAh/g decreased to 240 mAh/g after 50 cycles for graphite electrodes with 81% retention capacity. But the FSG paper itself as anode provides huge irreversible discharge capacity, i.e., 680 mAh/g at initial cycle dropped to 84 mAh/g second cycle. The retention capacity is very poor compared to graphite electrode and therefore it is concluded to be not a suitable candidate for anode material [48]. This helps infer that solid electrolyte interface (SEI) formation is a significant parameter to reduce the storage capacity in FSG electrodes.

To potentially apply FSG as anode material in LIBs, the second phase material with highly electrochemical active sites should be composited to enhance the capacity. In this regard, Lee et al. composited Si NPs on GO sheets, vacuum filtered, and followed by thermal treatment to produce FSG/Si nanoparticle (NP) paper. This work delivers high Li ion storage when compared to pristine FSG electrodes. Si NPs intercalated between the graphene sheets of FSG paper that facilitates good 3D graphite-like framework and provides high Li ion storage even at high current density [49]. Another work has been reported with similar hybrid FSG/Si NPs, whereas a facile route has been introduced to fabricate. The specific capacity of 708 mAh/g was observed without any loss even after 100 cycles and this is mainly due to the larger volume change in graphene-Si composite. It also denotes the performance of device with an efficient electron and charge transfer contributed by graphene sheets that minimize the internal resistance of the electrodes [50]. Zhang et al. prepared Si hollow nanosheets using Mg as template and connected with graphene sheets to obtain free-standing electrodes by layer-by-layer method followed by HI reduction treatment. The specific capacity was examined during flat and bent state, which delivers similar results without any loss. Remarkably, Si/FSG paper anodes retain high reversible capacities even at long cycles, which reveals their retention capacity. They exhibit specific capacity of 660 mAh/g at 0.2 A/g current density after 150 cycles with 99% coulombic efficiency [51]. As mentioned earlier, all the Si NPs are highly expensive in terms of manufacturing process and hence a low cost method plays a significant factor. To tackle this issue, Cai et al. prepared Si NPs on CNT surface using low-cost Al-Si alloy as starting material and further inserted with graphene sheets to form a self-standing hybrid anodes for LIBs. Comparing with bare Si/CNT or Si/Graphene anodes, Si-CNT/FSG hybrid electrode, it delivers 1100 mAh/g at 0.2 A/g current density after 100 cycles. Addition of CNT was involved to disperse the Si NPs on the surface and provide network between the graphene sheets for conductivity enhancement as well as improved Li ion intercalation for efficient charge transfer [52].

Metal oxides (MOs) play an important role in LIBs as anode material and their poor conductivity restricts their application. Hence, introducing the conductive phase into MOs provides high retention capacity with long-life cycling stability. The theoretical reversible capacity of SnO_2 is 782 mAh/g and its poor performance is due to low cycling with serious volume expansion. With this regard, SnO_2 NPs dispersed on GO surface, followed by vacuum filtration to obtain free-standing electrodes and used as two different LIB anodes by thermally reduced and chemically reduced respectively [53, 54]. The specific capacity of 438.5 mAh/g at 0.1 A/g and 700 mAh/g at 0.2 A/g has been delivered for the two different reduction methods for SnO_2 NPs/FSG electrodes. In both the cases, capacity fading is not observed even after the 50 cycles owing to the good anchoring of SnO_2 and graphene sheets. Further, other metal oxides TiO_2, Mn_3O_4, Fe_3O_4, and CuO nanostructured materials are incorporated into the FSG and are investigated for their performance in anode application for LIBs that delivers 269 mAh/g at 0.2 A/g, 692 mAh/g at 0.05 A/g, 544 mAh/g at 10 A/g, and 698.7 mAh/g at 0.67 A/g capacities, respectively [55–59]. Commonly, all these metal oxides' specific capacity shows a reasonable capacity with the long-life cycling after incorporating the MOs into FSG electrodes due to the following aspects: (1) Interaction of GO and MO precursors increases, which enhances the well dispersive growth of MO NPs on graphene sheets. (2) Anchoring of MOs and graphene enhances the volume expansion/contraction for lithiation/delithiation process. (3) The cycling stability increases compared to pristine MO anodes even after several cycles owing to its structural phase remain stable after alloying/de-alloying process of lithium ions. (4) MOs avoid the aggregation of graphene stacking that leads to larger void space to penetrate the electrolyte and make a strong interface with the electrochemical active MOs for an efficient Li ion storage.

Further, with the controlled synthesis of oxygen, functionalized CNT/FSG electrodes were fabricated for anode application in LIBs. The battery performance is based on the oxygen functional groups in the electrodes that have been investigated. An optimization in weight ratios of CNT/FSG and heat treatment improves the volumetric and gravimetric capacitances. The CNT/GO hybrid at a ratio of 1:1 shows higher volumetric capacity of 260 mAh/cm^3 that reduced at 200°C, while lower capacity of 43 mAh/cm^3 for 900°C treated CNT/GO. Whereas, at high current densities, the role of oxygen in capacity role suppress for 200°C larger than the 900°C [60]. This implies the importance of CNT intercalation between the graphene sheets of FSG electrodes. Zhang et al. demonstrated the defect-rich MoS_2 NSs/graphene/CNT hybrid paper as anode material for LIBs. In this design, MoS_2 facilitates the lithium ion storage due to the high active sites at the edges and the electrical conductivity improved by the network of CNTs attached to the graphene sheets. In addition to the conductivity enhancement, the porosity of the FSG electrodes increased by the network of CNT sandwiched graphene sheets. On the whole, the binder-free and substrate-free hybrid anode papers deliver high reversible capacity of 1137.2 mAh/g at 0.1 A/g current density with good cycling stability [61]. This framework induces a novel pathway to incorporate other host materials to understand the CNT/FSG electrodes. Recently, several transition metal oxides provide high reversible theoretical capacities compared with the commercialized graphite anode. To the CNT/FSG electrode network, transition metal oxides such as Fe_2O_3 [62], CuO [63], MnO [64], and $CoSnO_3$ [65] were incorporated as electrochemical active phase into the framework and investigated as anode material performance for LIBs. All these hybrid papers exhibit high reversible capacity of 716 and 600 mAh/g at 0.5 A/g current density more than 50 cycles for Fe_2O_3 and CuO nanobox, respectively. Apart from this, an enhanced capacity was observed for $CoSnO_3$ and MnO NPs at high current density of 2 A/g, which delivers 676 and 530 mAh/g, respectively. Individually, the CNT/FSG and transition metal oxide anodes were found to have a drastic decrease of specific capacity upon increasing the current density, whereas a slight decrease of specific capacity was observed after hosting the metal oxides into CNT/FSG framework. Reasons for high reversible capacity and good cyclic stability of metal oxide-CNT/FSG electrodes are very similar due to the following merits: (1) incorporation of metal oxides improves the Li ion kinetics and enhances the charge transfer due to highly conductive CNT network between the graphene sheets; (2) 3D framework of CNT/FSG has highly porous nature, large specific surface area, and large volume change, which has well dispersion of metal oxide NPs onto the carbon surfaces; and (3) long cycling due to good attachment of metal oxide with CNT/FSG, whereas greater the volume expansion, higher the Li ion intercalation.

Interestingly, Cao et al. designed a unique layered nanostructure of porous ternary $ZnCo_2O_4$ on graphene sheets and fabricated as flexible anode and investigated its electrochemical performance. And also they constructed full cell with $LiFePO_4$ as cathode material that deposited on FSG paper as slurry by homogenous mixing of conductive additive and polymer binder [66]. **Figure 4a** shows the photograph of flexible Li-ion battery fabricated by FSG hybrid electrodes. The half-cell of ZnCo2O4/FSG anode delivers higher specific capacity of 791 mAh/g at 1 A/g after 1000 cycles with 97.3% of capacity retention and concludes that it has an excellent cycling stability. **Figure 4b** shows the rate capability of the flexible battery with different current densities ranging from 0.5 to 10 C. This full cell delivers 40 mAh/g even at 10 C rate and the specific capacitance remains the same after the current density decreased to 2 C, which shows a good reversibility. The full cell has FSG paper as current collector for both the anode and cathode that are composited with $ZnCo_2O_4$ and $LiFePO_4$ as host materials, respectively. It operates at 2 V with initial charge of 143 mAh/g and coulombic efficiency of 97.2%, which is comparable to existing LIB. The specific capacity is maintained at 90 mAh/g with high capacity retention under flat and bent states over 100 cycling process, which implies the flexibility of the device as shown in **Figure 4d**. It represents that graphene conductivity is unchanged while bending the device.

4.2 Sodium-ion battery

Ahead of LIBs, SIBs have attracted the research community as the resources of Na are inexhaustible across the globe. In comparison with LIBs, the redox potential is −2.71 V vs. SHE and only the radius is 55% larger than the Li ions. Larger radius influences to focus on suitable material for insertion/extraction of Na ions effectively. The researchers focused on developing an efficient anode material for SIBs that involves carbon-based families and Na intermetallic compounds. The first cycle-specific capacity of sodium-antimony and sodium-phosphorous shows 600 and 2596 mAh/g,

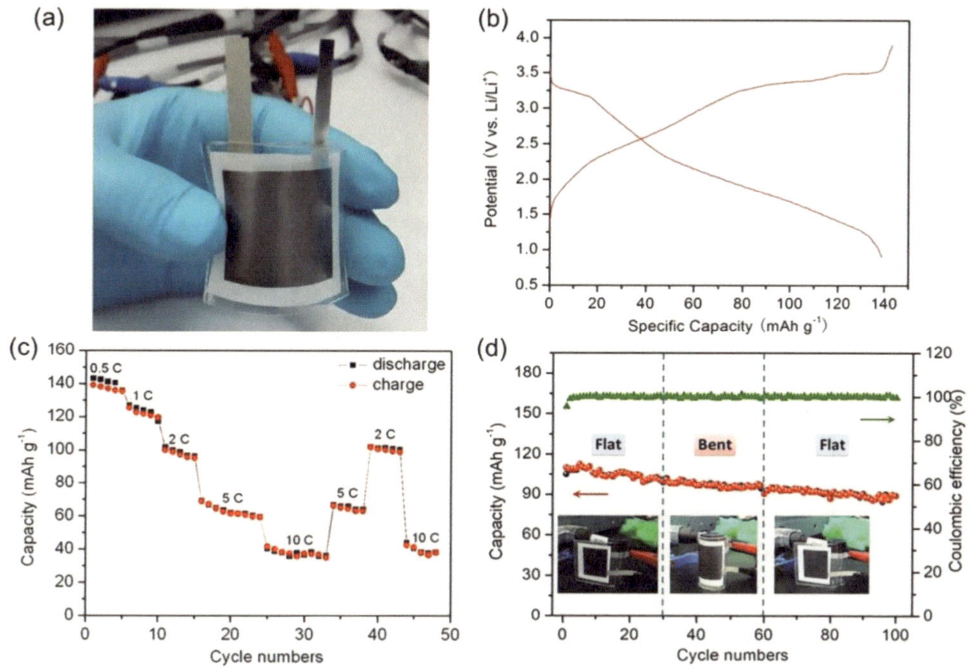

Figure 4.
(a) Photograph of flexible full cell Li-ion battery with FSG/$ZnCo_2O_4$ as anode and FSG/$LiFePO_4$ as cathode, (b) charge-discharge curve of full cell at 0.5 C rate, (c) charge-discharge rate capability at different rates, and (d) capacity variation on flat and bent state during cycling at 2 C rate (without permission from Ref. [66]).

respectively [67–69]. Specific capacities drop after first cycles due to the internal cracking in the electrodes upon Na ion insertion. It leads to hinder the electrical properties and dissolution of electrode materials to electrolyte. The hard carbon with large interlayer distance that functions as anode material for SIBs and delivers more than 200 mAh/g of capacity even after 100 cycles was reported elsewhere.

The porous nature and structure of the FSG could facilitate the accommodation of host materials such as transition metal chalcogenides (TMCs), which are electrochemically active for the Na ions for alloying process. David et al. reported that the MoS_2/FSG composite papers exhibit an excellent cyclic stability with high reversible capacity of 338 mAh/g at 0.025 A/g. It is the first report and opens the pathway to apply free-standing electrodes for SIB anode [70]. The cyclic stability was enhanced in flower-like MoS_2 incorporated on graphene foam prepared by one-step microwave-assisted synthesis. It offers stable capacity of 290 mAh/g at 0.1 A/g after 50 cycles compared to previous MoS_2/FSG electrode. The cycling performance is enhanced due to highly conductive 3D graphene foam and well-dispersed MoS_2, which shields as well as avoids the strain during the sodiation/desodiation process at anode [71]. With the significance of MoS_2 TMC for SIB anodes, further investigation was followed by incorporating other TMCs such as WS_2 and $Co_{0.85}Se$ into FSG [72, 73]. As mentioned in LIBs, the electrochemical behavior can be increased by introducing the heteroatoms into the graphene sheets. Heteroatom-doped FSG electrode performance was investigated for SIB anode, where the nitrogen improves the electronic conductivity and fluorine expands the interlayer for an efficient accommodation of Na ions. This delivers a reversible capacity of 56.3 mAh/g at 1 A/g for 5000 cycles. It indicates that the doping of heteroatoms enhances the cycling stability of SIB anodes. **Figure 5a** shows the discharge/charge profile before and after the bent state, which remains with the same capacity at current density

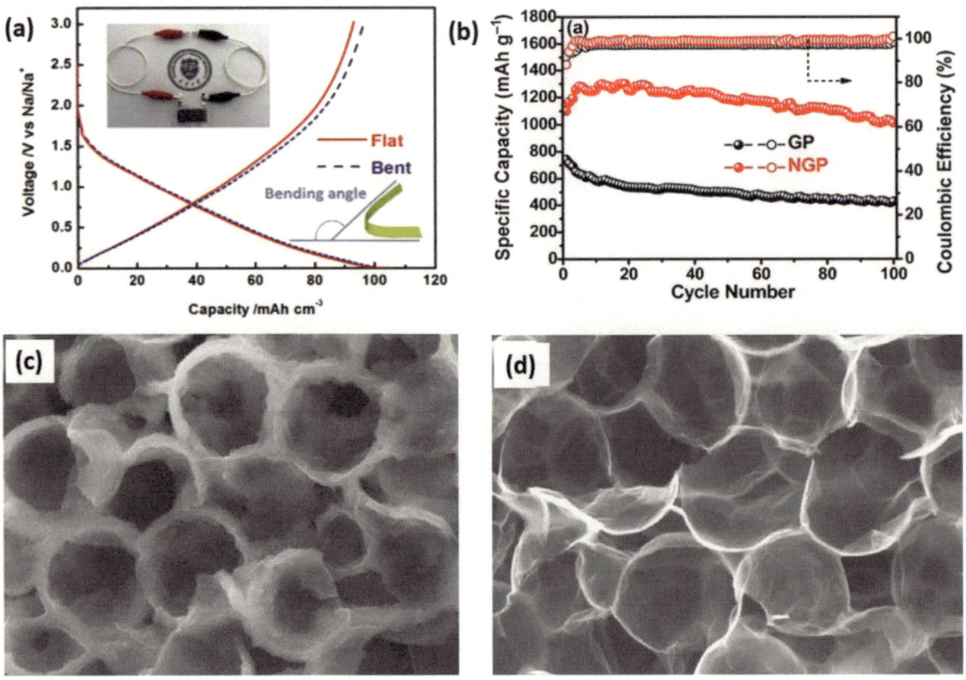

Figure 5.
(a) Discharge/charge profile of heteroatoms (N and F)-doped FSG electrode at bent and normal state for SIBs. (inset: The photograph of FSG pouch cell illuminated with LED), (b) comparison of specific capacity and coulombic efficiency of bare FSG and N-doped FSG for Li-S battery. Cross-sectional SEM images of (c) discharged and (d) re-charged macroporous FSG electrodes (without permission from Refs. [74–76]).

of 0.05 A/g. It reveals the mechanical strength of the FSG electrodes that is suitable to fabricate flexible pouch cell [74]. Even though the above said materials show an excellent cyclic stability, still it is necessary to improve the specific capacity of SIBs. It is well known that Na_3P has theoretical capacity of 2600 mAh/g, where its demerits are very similar to those of Si electrode in LIBs. Because of high pulverization, fast capacity fading and also it hinders the electrical contact which lags behind in the electrochemical stability. Lots of effort have been made by assembling red P into carbon matrix to overcome these problems. Red P was composited on carbon nanofibers (CNFs) and dipped in GO solution followed by HI treatment providing P-CNF/FSG electrodes. In this architecture, CNF network enhances the pathway of electron transport rapidly and the role of graphene sheets to improve the conductivity as well as to avoid the breakup of bonds P–P from electrodes. This work demonstrates a significant capacity of 406.6 mAh/g at 1 A/g after 180 cycles [77]. Moreover, the graphene sheets have been utilized as a multifunctional conductive binder, and hard carbon/FSG as anodes for SIBs was constructed. It delivers high reversible capacity of 372.4 mAh/g and shows capacity retention of 90% over 200 cycling. A superior performance is observed in the absence of PVDF binder with higher rate capabilities and converting the rigid nature of hard carbon into flexible graphene sheets [78].

4.3 Li-S battery

Akin to SIBs, FSG electrodes play a major role in other rechargeable secondary batteries such as Li-S, Li-air, and Zn-air. The higher specific energy is a significant parameter for transportation and stationary applications, and in that case, Li-S batteries offer advantages but it is limited with few challenges discussed later. The highest theoretical capacity of Li-S system is 2600 Wh/kg, which is highest than the LIB due to highest capacity of Li-S cathode sulfur has 1675 mAh/g. The most challenging part is to improve the electronic conductivity of cathodes of Li-S as the sulfur exhibits poor conductivity of 10–17 S/cm as well as the formation of polysulfides at cathodes. These polysulfides oxidize the Li anode and get back to cathodes and re-oxidize, thus lowering the performance of Li-S system. An extensive effort has been made to improve the cathodes by incorporating the carbon additives to sulfur to minimize the unnecessary reactions. Initially, mesoporous FSG was prepared and the sulfur was deposited by vapor treatment and was utilized as cathodes for Li-S system. It delivers charging capacity of 1288 mAh/g with high coulombic efficiency that reveals the restriction of sulfur to dissolute polysulfides in mesoporous FSG framework [79]. Similar to LIB and SIBs, the electrochemical behavior of cathode in Li-S system enhanced for heteroatom-doped FSG electrodes. **Figure 5b** shows the comparison of FSG and N-doped FSG capacity and coulombic efficiency with different cycle number. The heteroatom-doped FSG shows superior performance than the bare FSG due to the high interaction of polysulfides with heteroatoms that increase specific capacity. The nitrogen doping effect in FSG minimizes the concentration of polysulfides and forms a uniform layer of Li_2S at cathode. This system delivers 1000 mAh/g at 0.335 A/g after 100 cycles [75]. In another work, Zhu et al. developed free-standing cathodes by CNTs that were interconnected with the sulfur-graphene walls and investigated the electrochemical behavior that delivers 1346 mAh/g at 0.17 A/g current density. It is due to sulfur at graphene walls that deals to provide dual response as follows: (i) hinder the dissolution of polysulfides minimizing the shuttle phenomenon and (ii) offer volume expansion even at high quantity of sulfur. Moreover, its capacity retention shows 40% when current density is increased to 16.7 A/g owing to the good electron pathway by CNTs connected with graphene nanosheets [80]. Further, nanosized Li_2S (25–50 nm) particles incorporated into FSG papers by vacuum filtration process demonstrated an excellent cycling and rate capability with reversible capacity of 816.1 mAh/g at 0.1675 A/g (150 cycles) and 597 mAh/g at 11.7 A/g (200 cycles). This shows excellent performance in electrochemical behavior due to the uniform distribution of Li_2S particles on graphene sheets that minimize the barrier for Li ion transport and particularly it has superior wetting nature to interconnect the polysulfides with graphene network into the paper electrodes [81]. Similarly, Chen et al. designed an efficient hierarchical nanostructure like nanobundled forest with Li_2S/few-walled CNTs at FSG obtained solution processing followed by self-assembly method as cathodes. In this design, CNTs assembled in shaft-like structure and Li_2S as active material, whereas the graphene sheets act as barrier for Li_2S. It achieves high capacity of 868 and 433 mAh/g at current density of 335 and 16.7 A/g, respectively. This originates from the good framework between CNTs and graphene sheets as well as the uniform distribution of Li_2S, and moreover, the barrier of graphene sheets for Li_2S reduces the dissolution of polysulfides. Overall, the influence of void space enhances the volume change and thus improves the cycling stability of Li-S battery [82].

4.4 Metal-air battery

Recently, metal-air batteries have inspired much attention apart from the above said battery systems due to their high theoretical capacity than the metal-ion and Li-S batteries. The metal-air batteries can be operated in aqueous or nonaqueous medium based on the selection of metals. The nonaqueous medium is well suited for the Li-air batteries that deliver high capacity than in aqueous medium but still there are some issues when it comes to the practical application. The development of cathode in Li-air is significant as it is the main compartment to breathe oxygen for delivering high capacity of the system. There are a lot of reports for cathode development based on metal oxides grown on Ni foam as binder-free electrodes. The role of FSG electrodes was also investigated as cathodes for Li-air batteries. First, Kim et al. developed graphene nanoplates (GNP)/GO composite paper-like electrodes as cathodes for Li-air battery system. The wrinkled nature of the paper electrodes induces the high surface area and also delivers higher discharge capacity of 9760 mAh/g at 0.1 A/g current density. This superior performance is due to the reduced overpotential, and the difference in consumption/evolution of O_2 is minimized. On the whole, the system exhibits higher efficiency in OER (oxygen evolution reaction)/ORR (oxygen reduction reaction) of 87% [83]. The same group developed macroporous FSG paper with surface area of 373 m^2/g and pore volume of 10.9 cm^3/g with 91.6% of porosity that exhibits a high specific capacity of 12,200 mAh/g at 0.2 A/g. The rate capability is enhanced where it shows high cycling performance even at higher current density of 0.5 and 2 A/g that delivers approximately 1000 mAh/g. This is attributed to the minimized volume expansion that limits the decomposition and formation of Li_2O_2 at the macroporous nature of FSG. While discharging/charging the macroporous FSG, the nature of FSG electrode decomposes the discharge products completely that reveals its highly porous structure as shown in the **Figure 5c** and **d** [76]. Researchers investigated the effect of FSG cathodes in Li-air upon introduction of metal oxides, namely, α-MnO_2 and $NiCo_2O_4$. Upon insertion of α-MnO_2 into FSG electrodes, the overpotential decrease was caused during charge/discharge process. It delivers 2900 mAh/g for the higher content of α-MnO_2 that was reported and shows the catalytic improvement in this study [84]. And Jiang et al. reported an excellent reversible capacity of 5000 mAh/g at 0.4 A/g by incorporating mesoporous $NiCo_2O_4$ into macropores of FSG. It also lowers about 0.18 and 0.54 V of overpotential for discharge and charge, respectively [85].

5. Conclusions

In this chapter, FSG electrodes in battery applications signify their potential advantages to the fabrication technology. The fabrication of FSG electrode is facile as well as it excludes some additives applied in conventional electrodes. At present, the electrode of spent batteries contains active materials, binder, and metal foil, which set hurdles for recycling process. Herein, the FSG hybrid electrodes provide good capacity and cycling for battery application without binder and metal current collector. This exclusion provides light weight and flexible batteries and also there is a pathway to discover a facile route to recover the materials from FSG hybrid–based spent batteries in future.

Acknowledgements

This work was supported by South China Normal University. F.C. thanks the support from Outstanding Young Scholar Project (8S0256), the Project of Blue Fire Plan (CXZJHZ201709), and the Scientific and Technological Plan of Guangdong Province (2018A050506078).

Conflict of interest

The authors declare that there is no conflict of interest.

Author details

Karthick Ramalingam and Fuming Chen*
Guangdong Provincial Key Laboratory of Quantum Engineering and Quantum Materials, Guangdong Engineering Technology Research Center of Efficient Green Energy and Environment Protection Materials, School of Physics and Telecommunication Engineering, South China Normal University, Guangzhou, P.R. China

*Address all correspondence to: fmchen@m.scnu.edu.cn

References

[1] Neelu C, Ru Shi L. Electrochemical technologies for energy storage and conversion. In: Ru Shi L, Lei Z, Xueliang S, Hansan L, Jiujun Z, editors. Electrochemical Technologies for Energy Storage and Conversion. Wiley; 2012. pp. 1-43. DOI: 10.1002/9783527639496

[2] Dai L, Chang DW, Jong-Beom B, Lu W. Carbon nanomaterials for advanced energy conversion and storage. Small. 2012;**8**:1130-1166. DOI: 10.1002/smll.201101594

[3] Hou J, Shao Y, Ellis MW, Moore RB, Yi B. Graphene-based electrochemical energy conversion and storage: Fuel cells, supercapacitors and lithium ion batteries. Physical Chemistry Chemical Physics. 2011;**13**:15384-15402. DOI: 10.1039/C1CP21915D

[4] Liu J, Xue Y, Zhang M, Dai L. Graphene-based materials for energy applications. MRS Bulletin. 2012;**37**:1265-1272. DOI: 10.1557/mrs.2012.179

[5] Chen T, Dai L. Carbon nanomaterials for high-performance supercapacitors. Materials Today. 2013;**16**:272-280. DOI: 10.1016/j.mattod.2013.07.002

[6] Choi HJ, Jung SM, Seo JM, Chang DW, Dai L, Baek JB. Graphene for energy conversion and storage in fuel cells and supercapacitors. Nano Energy. 2012;**1**:534-551. DOI: 10.1016/j.nanoen.2012.05.001

[7] Feng L, Yang L, Huang Z, Luo J, Li M, Wang D, et al. Enhancing electrocatalytic oxygen reduction on nitrogen-doped graphene by active sites implantation. Scientific Reports. 2013;**3**:3306(1-8). DOI: 10.1038/srep03306

[8] Zhai P, Lee CC, Chang YH, Liu C, Wei TC, Feng SP. A significant improvement in the electrocatalytic stability of N-doped graphene nanosheets used as a counter electrode for $[Co(bpy)^{3}]^{3+/2+}$ based porphyrin-sensitized solar cells. ACS Applied Materials and Interfaces. 2015;**7**:2116-2123. DOI: 10.1021/am5083705

[9] Cai X, Lai L, Shen Z, Lin J. Graphene and graphene-based composites as Li-ion battery electrode materials and their application in full cells. Journal of Materials Chemistry A. 2017;**5**:15423-15446. DOI: 10.1039/C7TA04354F

[10] Yuan H, Kong L, Li T, Zhang Q. A review of transition metal chalcogenide/graphene nanocomposites for energy storage and conversion. Chinese Chemical Letters. 2017;**28**:2180-2194

[11] Srivastava M, Singh J, Kuila T, Layek RK, Kim NH, Lee JH. Recent advances in graphene and its metal-oxide hybrid nanostructures for lithium-ion batteries. Nanoscale. 2015;**7**:4820-4868. DOI: 10.1039/C4NR07068B

[12] Han S, Wu D, Li S, Zhang F, Feng X. Porous graphene materials for advanced electrochemical energy storage and conversion devices. Advanced Materials. 2014;**26**:849-864. DOI: 10.1002/adma.201303115

[13] Geim AK, Novoselov KS. The rise of graphene. Nature Materials. 2007;**6**:183-191. DOI: 10.1038/nmat1849

[14] Wang X, Shi Y. Fabrication techniques of graphene nanostructures. In: Gang Z, Navin M, editors. Nanofabrication and its Application in Renewable Energy. RSC Publishing: London; 2014. pp. 1-30. DOI: 10.1039/9781782623380-00001

[15] Cao MS, Wang XX, Cao WQ, Yuan J. Ultrathin graphene: Electrical properties and highly efficient

electromagnetic interference shielding. Journal of Materials Chemistry C. 2015;**3**:6589-6599. DOI: 10.1039/C5TC01354B

[16] Allen MJ, Tung VC, Kaner RB. Honeycomb carbon: A review of graphene. Chemical Reviews. 2010;**110**:132-145. DOI: 10.1021/cr900070d

[17] Kuchta B,Firlej L, Mohammadhosseini A, Boulet P, Beckner M, Romanos J, et al. Hypothetical high-surface-area carbons with exceptional hydrogen storage capacities: Open carbon frameworks. Journal of the American Chemical Society. 2012;**134**:15130-15137. DOI: 10.1021/ja306726u

[18] Frank IW, Tanenbaum DM, Van der Zande AM, McEuen PL. Mechanical properties of suspended graphene sheets. Journal of Vacuum Science and Technology, B: Microelectronics and Nanometer Structures—Processing, Measurement, and Phenomena. 2007;**25**:2558-2561. DOI: 10.1116/1.2789446

[19] Bizao RA, Botari T, Perim E, Nicola MP, Galvao DS. Mechanical properties and fracture patterns of graphene (graphitic) nanowiggles. Carbon. 2017;**119**:431-437. DOI: 10.1016/j.carbon.2017.04.018

[20] Brownson DAC, Banks CE. The Handbook of Graphene Electrochemistry. 1st ed. London: Springer-Verlag; 2014 . 174 p. DOI: 10.1007/978-1-4471-6428-9

[21] Bianco A, Cheng HM, Enoki T, Gogotsi Y, Hurt RH, Koratkar N, et al. All in the graphene family—A recommended nomenclature for two-dimensional carbon materials. Carbon. 2013;**65**:1-6. DOI: 10.1016/j.carbon.2013.08.038

[22] Zhang J, Hu P, Wang X, Wang Z, Liu D, Yang B, et al. CVD growth of large area and uniform graphene on tilted copper foil for high performance flexible transparent conductive film. Journal of Materials Chemistry. 2012;**22**:18283-18290. DOI: 10.1039/C2JM33881E

[23] Liu Z, Xie Y, Zhao J, Wu S, Zhou Z, Wen M, et al. Rapid preparation of conductive transparent films via solution printing of graphene precursor. Thin Solid Films. 2018;**657**:24-31. DOI: 10.1016/j.tsf.2018.05.005

[24] Karthick R, Brindha M, Selvaraj M, Ramu S. Stable colloidal dispersion of functionalized reduced graphene oxide in aqueous medium for transparent conductive film. Journal of Colloid and Interface Science. 2013;**406**:69-74. DOI: 10.1016/j.jcis.2013.06.006

[25] Ambrosi A, Chua CK, Latiff NM, Loo AH, Wong CHA, Eng AYS, et al. Graphene and its electrochemistry—An update. Chemical Society Reviews. 2016;**45**:2458-2493. DOI: 10.1039/C6CS00136J

[26] Li D, Muller MB, Gilje S, Kaner RB, Wallace GG. Processable aqueous dispersions of graphene nanosheets. Nature Nanotechnology. 2008;**3**:101-105. DOI: 10.1038/nnano.2007.451

[27] Sun L, Fungetsu B. Mass production of graphene oxide from expanded graphite. Materials Letters. 2013;**109**:207-210. DOI: 10.1016/j.matlet.2013.07.072

[28] Konios D, Stylianakis M, Stratakis E, Kymakis E. Dispersion behaviour of graphene oxide and reduced graphene oxide. Journal of Colloid and Interface Science. 2014;**430**:108-112. DOI: 10.1016/j.jcis.2014.05.033

[29] Rinzler AG, Liu J, Dai H, Nikolaev P, Huffman CB, Rodríguez-Macías FJ, et al. Large-scale purification of single-wall carbon nanotubes: Process, product, and characterization. Applied

Physics A. 1998;**67**:29-37. DOI: 10.1007/s0033900507

[30] Ding W, Pengcheng S, Changhong L, Wei W, Shoushan F. Highly oriented carbon nanotube papers made of aligned carbon nanotubes. Nanotechnology. 2008;**19**:075609. DOI: 10.1088/0957-4484/19/7/075609

[31] Dikin DA, Stankovich S, Zimney EJ, Piner RD, Dommett GHB, Evmenenko G, et al. Preparation and characterization of graphene oxide paper. Nature. 2007;**448**:457-460. DOI: 10.1038/nature06016

[32] Pei S, Zhao J, Du J, Ren W, Cheng HM. Direct reduction of graphene oxide films into highly conductive and flexible graphene films by hydrohalic acids. Carbon. 2010;**48**:4466-4474. DOI: 10.1016/j.carbon.2010.08.006

[33] Liu C, Hao F, Zhao X, Zhao Q, Luo S, Lin H. Low temperature reduction of free-standing graphene oxide papers with metal iodides for ultrahigh bulk conductivity. Scientific Reports. 2014;**4**:3965. DOI: 10.1038/srep03965

[34] Chen C, Yang QH, Yang Y, Lu W, Wen Y, Hou PX, et al. Self-assembled free-standing graphite oxide membrane. Advanced Materials. 2009;**21**:3007-3011. DOI: 10.1002/adma.200803726

[35] Huang ZD, Zhang B, Liang R, Zheng QB, Oh SW, Lin XY, et al. Effects of reduction process and carbon nanotube content on supercapacitive performance of flexible graphene oxide papers. Carbon. 2012;**50**:4239-4251. DOI: 10.1016/j.carbon.2012.05.006

[36] Park S, Lee KS, Bozoku G, Cai W, Nguyen SBT, Ruoff RS. Graphene oxide papers modified by divalent ions—Enhancing mechanical properties via chemical cross-linking. ACS Nano. 2008;**2**:572-578. DOI: 10.1021/nn700349a

[37] Tan Y, Song Y, Zheng Q. Facile regulation of glutaraldehyde-modified graphene oxide for preparing free-standing papers and nanocomposite films. Chinese Journal of Polymer Science. 2013;**31**:399-406. DOI: 10.1007/s10118-013-1234-7

[38] Marquez-Lamas U, Martinez-Guerra E, Toxqui-Teran A, Aguirre-Tostado FS, Lara-Ceniceros TE, Bonilla-Cruz J. Tuning mechanical, electrical and optical properties of flexible and free-standing functionalized graphene oxide papers having different interlayer d-spacing. The Journal of Physical Chemistry C. 2017;**121**:852-862. DOI: 10.1021/acs.jpcc.6b09961

[39] Medhekar NV, Ramasubramaniam A, Ruoff RS, Shenoy VB. Hydrogen bond networks in graphene oxide composite paper: Structure and mechanical properties. ACS Nano. 2010;**4**:2300-2306. DOI: 10.1021/nn901934u

[40] Alzate-Carvajal N, Acevedo-Guzman DA, Meza-Laguna V, Farıas MH, Perez-Rey LA, Abarca-Morales E, et al. One-step nondestructive functionalization of graphene oxide paper with amines. RSC Advances. 2018;**8**:15253-15265. DOI: 10.1039/c8ra00986d

[41] Vallés C, David NJ, Benito AM, Maser WK. Flexible conductive graphene paper obtained by direct and gentle annealing of graphene oxide paper. Carbon. 2012;**50**:835-844. DOI: 10.1016/j.carbon.2011.09.042

[42] Gao J, Liu C, Miao L, Wang X, Chen Y. Free-standing reduced graphene oxide paper with high electrical conductivity. Journal of Electronic Materials. 2016;**45**:1290-1295. DOI: 10.1007/s11664-015-4000-5

[43] Tarascon JM, Armand M. Issues and challenges facing rechargeable lithium

batteries. Nature. 2001;**414**:359-367. DOI: 10.1038/35104644

[44] Hadjipaschalis I, Poullikkas A, Efthimiou V. Overview of current and future energy storage technologies for electric power applications. Renewable and Sustainable Energy Reviews. 2009;**13**:1513-1522. DOI: 10.1016/j.rser.2008.09.028

[45] Younesi R, Veith GM, Johansson P, Edström K, Vegge T. Lithium salts for advanced lithium batteries: Li–metal, Li–O_2, and Li–S. Energy and Environmental Science. 2015;**8**:1905-1922. DOI: 10.1039/C5EE01215E

[46] Ng SH, Wang J, Guo ZP, Chen J, Wang GX, Liu HK. Single wall carbon nanotube paper as anode for lithium-ion battery. Electrochimica Acta. 2005;**51**:23-28. DOI: 10.1016/j.electacta.2005.04.045

[47] Chen J, Minett AI, Liu Y, Lynam C, Sherrell P, Wang C, et al. Direct growth of flexible carbon nanotube electrodes. Advanced Materials. 2008;**20**:566-570. DOI: 10.1002/adma.200701146

[48] Wang C, Li D, Too CO, Wallace GG. Electrochemical properties of graphene paper electrodes used in lithium batteries. Chemistry of Materials. 2009;**21**: 2604-2606. DOI: 10.1021/cm900764n

[49] Lee JK, Smith KB, Hayner CM, Kung HH. Silicon nanoparticles–graphene paper composites for Li ion battery anodes. Chemical Communications. 2010;**46**:2025-2027. DOI: 10.1039/b919738a

[50] Wang JZ, Zhong C, Chou SL, Liu HK. Flexible free-standing graphene-silicon composite film for lithium-ion batteries. Electrochemistry Communications. 2010;**12**:1467-1470. DOI: 10.1016/j.elecom.2010.08.008

[51] Zhang H, Jing S, Hu Y, Jiang H, Li C. A flexible freestanding Si/rGO hybrid film anode for stable Li-ion batteries. Journal of Power Sources. 2016;**307**:214-219. DOI: 10.1016/j.jpowsour.2015.12.107

[52] Cai H, Han K, Jiang H, Wang J, Liu H. Self-standing silicon-carbon nanotube/graphene by a scalable in situ approach from low-cost Al-Si alloy powder for lithium ion batteries. Journal of Physics and Chemistry of Solids. 2017;**109**:9-17. DOI: 10.1016/j.jpcs.2017.05.009

[53] Liang J, Zhao Y, Guo L, Li L. Flexible free-standing graphene/SnO_2 nanocomposites paper for Li-ion battery. ACS Applied Materials and Interfaces. 2012;**4**:5742-5748. DOI: 10.1021/am301962d

[54] Gao T, Huang K, Qi X, Li H, Yang L, Zhong J. Free-standing SnO_2 nanoparticles@graphene hybrid paper for advanced lithium-ion batteries. Ceramics International. 2014;**40**:6891-6897. DOI: 10.1016/j.ceramint.2013.12.009

[55] Hu T, Sun X, Sun H, Yu M, Lu F, Liu C, et al. Flexible free-standing graphene–TiO_2 hybrid paper for use as lithium ion battery anode materials. Carbon. 2013;**51**:322-326. DOI: 10.1016/j.carbon.2012.08.059

[56] Ren HM, Ding YH, Chang FH, He X, Feng JQ, Wang CF, et al. Flexible free-standing TiO_2/graphene/PVdF films as anode materials for lithium-ion batteries. Applied Surface Science. 2012;**263**:54-57. DOI: 10.1016/j.apsusc.2012.08.107

[57] Park SK, Seong CY, Yoo S, Piao Y. Porous Mn_3O_4 nanorod/reduced graphene oxide hybrid paper as a flexible and binder-free anode material for lithium ion battery. Energy. 2016;**99**:266-273. DOI: 10.1016/j.energy.2016.01.061

[58] Guo J, Zhu H, Sun Y, Tang L, Zhang X. Pie-like free-standing paper of

graphene paper@Fe_3O_4 nanorod array@ carbon as integrated anode for robust lithium storage. Chemical Engineering Journal. 2017;**309**:272-277. DOI: 10.1016/j.cej.2016.10.041

[59] Liu Y, Wang W, Gu L, Wang Y, Ying Y, Mao Y, et al. Flexible CuO nanosheets/reduced-graphene oxide composite paper: Binder-free anode for high performance lithium-ion batteries. ACS Applied Materials and Interfaces. 2013;**5**:9850-9855. DOI: 10.1021/am403136e

[60] Byon HR, Gallant BM, Lee SW, Yang SH. Role of oxygen functional groups in carbon nanotube/graphene freestanding electrodes for high performance lithium batteries. Advanced Functional Materials. 2013;**23**:1037-1045. DOI: 10.1002/adfm.201200697

[61] Zhang L, Fan W, Liu T. A flexible free-standing defect-rich MoS_2/graphene/carbon nanotube hybrid paper as a binder-free anode for high-performance lithium ion batteries. RSC Advances. 2015;**5**:43130-43140. DOI: 10.1039/c5ra05038c

[62] Wang J, Wang G, Wang H. Flexible free-standing Fe_2O_3/graphene/carbon nanotubes hybrid films as anode materials for high performance lithium-ion batteries. Electrochimica Acta. 2015;**182**:192-201. DOI: 10.1016/j.electacta.2015.09.080

[63] Liu Y, Cai X, Shi W. Free-standing graphene/carbon nanotubes/CuO aerogel paper anode for lithium ion batteries. Materials Letters. 2016;**172**:72-75. DOI: 10.1016/j.matlet.2016.02.068

[64] Li Y, Wang P, Bao Y, Huang K. A flexible nanostructured paper of MnO NPs@MWCNTs/r-GO multilayer sandwich composite for high-performance lithium-ion batteries. Ceramics International. 2017;**43**:7588-7593. DOI: 10.1016/j.ceramint.2017.03.051

[65] Zhao X, Wang G, Zhou Y, Wang H. Flexible free-standing ternary $CoSnO_3$/graphene/carbon nanotubes composite papers as anodes for enhanced performance of lithium-ion batteries. Energy. 2017;**118**:172-180. DOI: 10.1016/j.energy.2016.12.018

[66] Cao H, Zhou X, Deng W, Ma Z, Liu Y, Liu Z. Layer structured graphene/porous $ZnCo_2O_4$ composite film for high performance flexible lithium-ion batteries. Chemical Engineering Journal. 2018;**343**:654-661. DOI: 10.1016/j.cej.2018.03.001

[67] Qian J, Chen Y, Wu L, Cao Y, Ai X, Yang H. High capacity Na-storage and superior cyclability of nanocomposite Sb/C anode for Na-ion batteries. Chemical Communications. 2012;**48**:7070-7072

[68] Darwiche A,Marino C, Sougrati MT, Fraisse B, Stievano L, Monconduit L. Better cycling performances of bulk Sb in Na-ion batteries compared to Li-ion systems: An unexpected electrochemical mechanism. Journal of the American Chemical Society. 2012;**134**:20805-20811. DOI: 10.1021/ja310347x

[69] Qian J, Wu X, Cao Y, Ai X, Yang H. High capacity and rate capability of amorphous phosphorus for sodium ion batteries. Angewandte Chemie, International Edition. 2013;**52**:4633-4636. DOI: 10.1002/anie.201209689

[70] David L, Bhandavat R, Singh G. MoS_2/Graphene composite paper for sodium-ion battery electrodes. ACS Nano. 2014;**8**:1759-1770. DOI: 10.1021/nn406156b

[71] Xiang J, Dong D, Wen F, Zhao J, Zhong X, Wang L, et al. Microwave

synthesized self-standing electrode of MoS_2 nanosheets assembled on graphene foam for high-performance Li-ion and Na-ion batteries. Journal of Alloys and Compounds. 2016;**660**:11-16. DOI: 10.1016/j.jallcom.2015.11.040

[72] Wang Y, Kong D, Shi W, Liu B, Sim G, Ge Q, et al. Ice Templated free-standing hierarchically WS_2/CNT-rGO aerogel for high-performance rechargeable lithium and sodium ion batteries. Advanced Energy Materials. 2016;**6**:1601057. DOI: 10.1002/aenm.201601057

[73] Zhang G, Liu K, Liu S, Song H, Zhou J. Flexible Co0.85Se nanosheets/graphene composite film as binder-free anode with high Li- and Na-ion storage performance. Journal of Alloys and Compounds. 2018;**731**:714-722. DOI: 10.1016/j.jallcom.2017.10.094

[74] An H, Li Y, Gao Y, Cao C, Han J, Feng Y, et al. Free-standing fluorine and nitrogen co-doped graphene paper as a high-performance electrode for flexible sodium-ion batteries. Carbon. 2017;**116**:338-346. DOI: 10.1016/j.carbon.2017.01.101

[75] Han K, Shen J, Hao S, Hao S, Ye H, Wolverton C, et al. Free-standing nitrogen-doped graphene paper as electrodes for high-performance lithium/dissolved polysulfide batteries. ChemSusChem. 2014;**7**:2545-2553. DOI: 10.1002/cssc.201402329

[76] Kim DY, Kim M, Kim DW, Suk J, Park JJ, Park OO, et al. Graphene paper with controlled pore structure for high-performance cathodes in Li_2O batteries. Carbon. 2016;**100**:265-272. DOI: 10.1016/j.carbon.2016.01.013

[77] Ma X, Chen L, Ren X, Hou G, Chen L, Zhang L, et al. High-performance red phosphorus/carbon nanofibers/graphene free-standing paper anode for sodium ion batteries. Journal of Materials Chemistry A. 2018;**6**:1574-1581. DOI: 10.1039/C7TA07762A

[78] Sun N, Guan Y, Liu YT, Zhu Q, Shen J, Liu S, et al. Facile synthesis of free-standing, flexible hard carbon anode for high performance sodium ion batteries using graphene as a multifunctional binder. Carbon. 2018;**137**:475-483. DOI: 10.1016/j.carbon.2018.05.056

[79] Huang X, Sun B, Li K, Chen S, Wang G. Mesoporous graphene paper immobilised sulfur as a flexible electrode for lithium-sulfur batteries. Journal of Materials Chemistry A. 2013;**1**:13484-13489. DOI: 10.1039/C3TA12826A

[80] Zhu L, Peng HJ, Liang J, Huang JQ, Chen CM, Guo X, et al. Interconnected carbon nanotube/graphene nanosphere scaffolds as free-standing paper electrode for high-rate and ultra-stable lithium-sulfur batteries. Nano Energy. 2015;**11**:746-755. DOI: 10.1016/j.nanoen.2014.11.062

[81] Wang C, Wang X, Yang Y, Kushima A, Chen J, Huang Y, et al. Slurryless Li_2S/reduced graphene oxide cathode paper for high-performance lithium sulfur battery. Nano Letters. 2015;**15**:1796-1802. DOI: 10.1021/acs.nanolett.5b00112

[82] Chen Y, Lu S, Zhou J, Qin W, Wu X. Synergistically assembled Li_2S/FWNTs@reduced graphene oxide nanobundle forest for free-standing high-performance Li_2S cathodes. Advanced Functional Materials. 2017;**27**:1700987. DOI: 10.1002/adfm.201700987

[83] Kim DY, Kim M, Kim DW, Suk J, Park OO, Kang Y. Flexible binder-free graphene paper cathodes for high-performance Li-O_2 batteries. Carbon. 2015;**93**:625-635. DOI: 10.1016/j.carbon.2015.05.097

[84] Ozcan S, Tokur M, Cetinkaya T, Guler A, Uysal M, Guler MO, et al. Free standing flexible graphene oxide $+\alpha$-MnO_2 composite cathodes for Li–air batteries. Solid State Ionics. 2016;**286**:34-39. DOI: 10.1016/j.ssi.2015.12.016

[85] Jiang Y, Zou L, Cheng J, Huang Y, Jia L, Chi B, et al. Needle-like $NiCo_2O_4$ coated on graphene foam as a flexible cathode for lithium-oxygen batteries. ChemElectroChem. 2017;**4**:3140-3147. DOI: 10.1002/celc.201700864

Chapter 7

Vanadium Redox Flow Batteries: Electrochemical Engineering

Sangwon Kim

Abstract

The importance of reliable energy storage system in large scale is increasing to replace fossil fuel power and nuclear power with renewable energy completely because of the fluctuation nature of renewable energy generation. The vanadium redox flow battery (VRFB) is one promising candidate in large-scale stationary energy storage system, which stores electric energy by changing the oxidation numbers of anolyte and catholyte through redox reaction. This chapter covers the basic principles of vanadium redox flow batteries, component technologies, flow configurations, operation strategies, and cost analysis. The thermodynamic analysis of the electrochemical reactions and the electrode reaction mechanisms in VRFB systems have been explained, and the analysis of VRFB performance according to the flow field and flow rate has been described. It is shown that the limiting current density of "flow-by" design is more than two times greater than that of "flow-through" design. In the cost analysis of 10 kW/120 kWh VRFB system, stack and electrolyte account for 40 and 32% of total cost, respectively.

Keywords: vanadium electrolyte, carbon electrode, overpotential, polarization, state of charge, flow-through, flow-by, flow rate, limiting current density, peak power density

1. Introduction

The global environmental is changing rapidly. The established world's first energy demand and biggest carbon emitter countries are being replaced by emerging countries. The use of renewable energy is expanding due to technological development and environmental problems. The global energy market is moving toward the reduction of fossil fuels and the expansion of environment friendly energy, a shift in the energy mix.

For stable supply of renewable energy with high volatility such as sunlight or wind power, securing stability of power system is the most important. To do this, an intelligent power network should be built up, and grid-based energy storage technology should be secured.

The vanadium redox flow battery is one of the most promising secondary batteries as a large-capacity energy storage device for storing renewable energy [1, 2, 4]. Recently, a safety issue has been arisen by frequent fire accident of a large-capacity energy storage system (ESS) using a lithium ion battery. The vanadium electrolyte is a nonflammable aqueous solution and has a high heat capacity to limit the temperature rise. Therefore, VRFB has no risk of ignition and explosion.

IntechOpen

The power of VRFB depends on the performance of the stack, and the energy storage capacity depends on the electrolyte concentration and the electrolyte reservoir size, which greatly increases the degree of freedom in system design [7, 24]. A schematic diagram of the vanadium redox flow battery is shown in **Figure 1**.

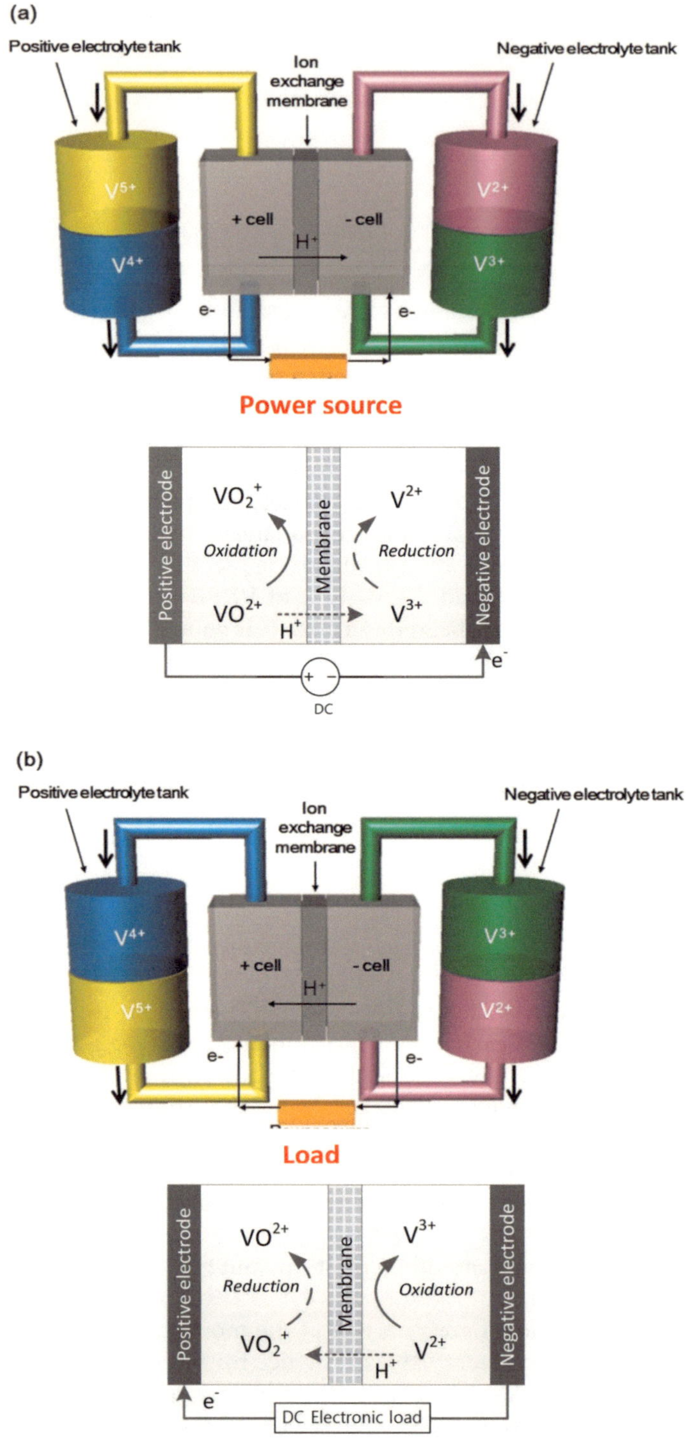

Figure 1.
Schematic of vanadium redox flow batteries: (a) charging and (b) discharging. Reproduced with permission from [3]. Copyright 2017 by Elsevier.

Flow batteries suffer from the capacity imbalance due to the mixing of the both side active materials caused by the electrolyte diffusion across the membrane, resulting in an irreversible loss of capacity as well as an efficiency loss [10–14]. Since the vanadium redox flow battery uses vanadium as the active material of both electrolytes, the use of appropriate rebalancing techniques can mitigate capacity loss though vanadium crossovers can lead to loss of efficiency.

2. Electrochemical reactions and kinetics

The vanadium ion may have various oxidation numbers from bivalent to pentavalent. Using this property, vanadium is used as the electrolyte redox couple material of the flow battery. $VO_2{}^+$, VO^{2+}, V^{3+}, and V^{2+} are represented by V(V), V(IV), V(III), and V(II) for explanation. Solution of V(III) is added to the negative electrolyte tank, and solution of V(IV) is added to the positive electrolyte tank as shown in **Figure 1**. When the electricity is applied to the electrodes, the V(III) ion of the negative electrolyte is reduced to V(II), and the V(IV) ion of the positive electrolyte is oxidized to V(V). This means that when the VRFB is charged, the difference in the oxidation number between the positive electrolyte and negative electrolyte increases from +1 to +3, and it can be understood conceptually that the electric energy is stored in the increased bivalent oxidation number. When the VRFB is discharged, V(II) in negative electrolyte is oxidized to V(III), and V(V) in positive electrolyte is reduced to V(IV). The chemical reactions for charge-discharge are expressed as follows:

$$\text{Negative electrode: } V^{2+} \leftrightarrow V^{3+} + e^- \quad E^0 = -0.255\ \text{V} \tag{1}$$

$$\text{Positive electrode: } VO_2^+ + e^- + 2H^+ \leftrightarrow VO^{2+} + H_2O\ E^0 = +1.004\ \text{V} \tag{2}$$

$$\text{Overall reaction: } VO_2^+ + V^{2+} + 2H^+ \leftrightarrow VO^{2+} + V^{3+} + H_2O\ E^0 = +1.259\ \text{V} \tag{3}$$

The permeation of the vanadium ions through the membrane occurs since any membrane cannot block the crossover of the redox species completely. The vanadium ions diffused to the counter electrolyte cause a cross-contamination reaction as below:

$$VO^{2+} + 2H^+ + e^- \leftrightarrow V^{3+} + H_2O\ E^0 = +0.34\ \text{V} \tag{4}$$

The self-discharging reactions caused by the vanadium ions permeated into the counter electrolytes can be described as below:

Negative electrode:

$$VO^{2+} + V^{2+} + 2H^+ \rightarrow 2V^{3+} + H_2O \tag{5}$$

$$VO_2^+ + 2V^{2+} + 4H^+ \rightarrow 3V^{2+} + 2H_2O \tag{6}$$

$$VO_2^+ + V^{3+} \rightarrow 2\ VO^{2+} \tag{7}$$

Positive electrode:

$$V^{2+} + 2VO_2^+ + 2H^+ \rightarrow 3VO^{2+} + H_2O \tag{8}$$

$$V^{3+} + VO_2^+ \rightarrow 2\ VO^{2+} \tag{9}$$

$$V^{2+} + VO^{2+} + 2H^+ \rightarrow 2\ V^{3+} + H_2O \tag{10}$$

When the VRFB is overcharged, hydrogen and oxygen gas can be generated at the negative and positive electrodes, respectively. Additionally, the carbon dioxide gas can be generated by corrosion of graphite plate with the produced oxygen gas.

Negative electrode:

$$2H^{+} + 2e^{-} \rightarrow H_{2(gas)} \tag{11}$$

Positive electrode:

$$C + 2H_2O \rightarrow CO_{2(gas)} + 4H^{+} + 4e^{-} \tag{12}$$

$$2H_2O \rightarrow O_{2(gas)} + 4H^{+} + 4e^{-} \tag{13}$$

$$C + O_{2(gas)} \rightarrow CO_{2(gas)} \tag{14}$$

The equilibrium cell potentials, E_{eq} for each reaction, are calculated using Nernst equation according to

$$E_{eq,\,neg} = E^0_{neg} + \frac{RT}{F} ln\left(\frac{C_{V^{3+}}}{C_{V^{2+}}}\right) \tag{15}$$

$$E_{eq,\,pos} = E^0_{pos} + \frac{RT}{F} ln\left(\frac{C_{VO_2^+}(c_{H^+})^2}{C_{VO^{2+}}}\right) \tag{16}$$

$$E_{eq,\,overall} = E^0_{overall} + \frac{RT}{F} ln\left(\frac{C_{VO_2^+}(C_{H^+})^2}{C_{VO^{2+}}}\frac{C_{V^{2+}}}{C_{V^{3+}}}\right) \tag{17}$$

where C_i^* is the concentration of the i species; E^0 is the standard cell potential for the electrode reaction; R is the ideal gas constant, 8.314 J/mol K; T is the cell temperature; and F is Faraday's constant, 96,485 As/mol.

The exchange current density is the magnitude of the current when the electrode reactions reach the equilibrium and can be described as

$$i_{0,\,neg} = Fk^0_{neg} C^{*\,(1-\alpha_{neg})}_{V^{3+}} C^{*\,\alpha_{neg}}_{V^{2+}} \tag{18}$$

$$i_{0,\,pos} = Fk^0_{pos} C^{*\,(1-\alpha_{pos})}_{VO_2^+} C^{*\,\alpha_{pos}}_{VO^{2+}} \tag{19}$$

where k^0 is the standard rate constant.

Following the Butler-Volmer equation [5, 24], the currents at negative electrode and positive electrode are described as

$$i_{neg} = i_{0,\,neg}\left[\left(\frac{C_{V^{3+}}(0,t)}{C^*_{V^{3+}}}\right)\exp\left(-\frac{\alpha_{neg}F}{RT}\eta_{neg}\right) - \left(\frac{C_{V^{2+}}(0,t)}{C^*_{V^{2+}}}\right)\exp\left(\frac{(1-\alpha_{neg})F}{RT}\eta_{neg}\right)\right] \tag{20}$$

$$i_{pos} = i_{0,\,pos}\left[\left(\frac{C_{VO_2^+}(0,t)}{C^*_{VO_2^+}}\right)\exp\left(-\frac{\alpha_{pos}F}{RT}\eta_{pos}\right) - \left(\frac{C_{VO^{2+}}(0,t)}{C^*_{VO^{2+}}}\right)\exp\left(\frac{(1-\alpha_{pos})F}{RT}\eta_{pos}\right)\right] \tag{21}$$

where α is the transfer coefficient or symmetry factor and η is the overpotential, defined as $\eta = \phi_s - \phi_l - E_{eq}$.

where φ_s is the electric potential of the solid electrode and φ_l is the electrolyte potential.

The standard open-circuit voltage of VRFB, $E^0 = 1.26$ V, can be derived from Gibbs free energy relation as below:

$$\Delta G^0 = \Delta H^0 - T\Delta S^0 = -nFE^0 = -119.3\ kJ/mol \tag{22}$$

However, the actual operating voltage of VRFB differs from this thermodynamic value. Charging voltage should be larger than 1.26 V since the amount of overpotential is required in addition to the thermodynamic voltage. **Figure 2** shows the relationship of the voltage and current during charging and discharging at the two electrodes of VRFB, assuming that the overall kinetics are determined by the charge transfer in the electrochemical reaction.

$$E_{charge} = E^0_{cell} + \eta_a + \eta_c + iR_{total} \tag{23}$$

$$E_{discharge} = E^0_{cell} - \eta_a - \eta_c - iR_{total} \tag{24}$$

where η_a is anodic overpotential and η_c is cathodic overpotential.

At discharge, the operating voltage becomes smaller than theoretical value. As the current density increases, the overpotential and iR drop increase, so the charging voltage increases and the discharging voltage decreases as shown in **Figure 3c**. Energy density and power density can be calculated in Eqs. (25) and (26), respectively.

$$\text{Energy density} = \frac{nCFV_{dis}}{N_{tank}} = \frac{1 \times \frac{1.6mol}{L} \times 26.8\frac{Ah}{mol} \times 1.3\ V}{2} = 27.872\ Wh/L \tag{25}$$

$$\text{Power denisty} = \text{current density} \times \overline{V}_{discharge} \tag{26}$$

where n is the number of electrons transferred during reactions, C is a vanadium electrolyte concentration, 1.6 mol/L, $\overline{V}_{discharge}$ is averaged discharge voltage, and

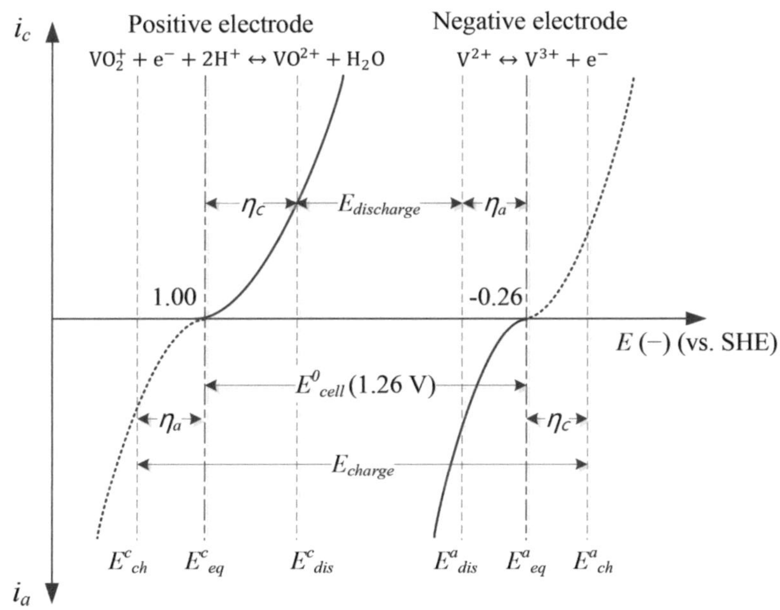

Figure 2.
Charge-discharge voltage of vanadium redox flow battery: Current vs. voltage and overpotential and open-circuit voltage at positive electrode and negative electrode.

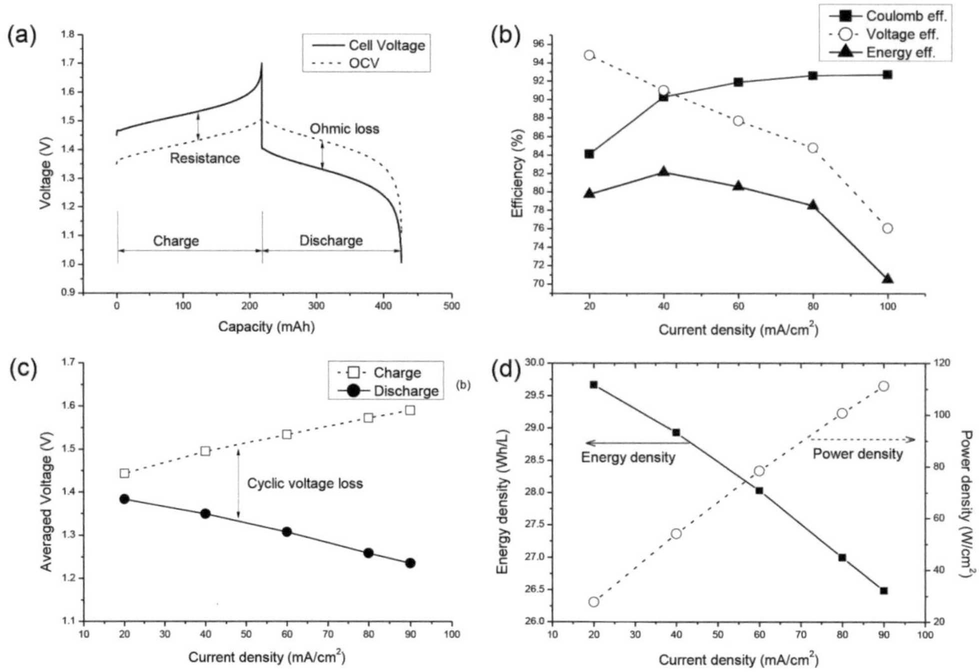

Figure 3.
Vanadium redox flow battery performance: (a) cell voltage and open-circuit voltage profiles at current density of 60 mA/cm², (b) efficiencies depending on current densities, (c) polarization plot of the unit cell, and (d) energy density and power density.

N_{tank} is a number of tank. There are only three variables that contribute to increasing energy density and power density: the vanadium ion concentration, discharging voltage, and current density. However, the concentration of the vanadium ions is limited by low solubilities of vanadium ions in aqueous solution. The discharging voltage and current density are restricted by the electrochemical activities of vanadium electrolyte. **Figure 3d** shows that as a current density increases, energy density decreases, and power density increases. Normal operating current density range is 50–80 mA/cm^2, and stored energy density is in the range of 25–35 Wh/L or 20–32 Wh/kg. The corresponding power density is less than 0.1 W/cm^2.

The performance of VRFB can be measured with three efficiencies: current efficiency, voltage efficiency, and energy efficiency, which are defined in Eqs. (27), (28), and (29), respectively. The current efficiency (CE, Coulombic efficiency) is defined as the ratio of the amount of usable charge to the stored charge amount, that is, the discharge capacity divided by the charge capacity. CE is a measure of the storage capacity loss during charge-discharge process. The capacity loss is mainly caused by the crossover of the electrolyte ions through the membrane. The mixed active materials result in a capacity imbalance between the anode and cathode electrolytes and an irreversible capacity loss.

$$CE = \frac{discharge\ capacity}{charge\ capacity} \times 100\% = \frac{\int_{dis} I(t)dt}{\int_{ch} I(t)dt} \times 100\% = \frac{I_{dis} \bullet t_{dis}}{I_{ch} \bullet t_{ch}} \times 100\% \\ = \frac{t_{dis}}{t_{ch}} \times 100\% (\text{If } I_{dis} = I_{ch}) \tag{27}$$

$$VE = \frac{average\ discharge\ voltage}{average\ charge\ voltage} \times 100\% = \frac{\int_{dis} V(t)dt / t_{dis}}{\int_{ch} V(t)dt / t_{ch}} \times 100\% \tag{28}$$

$$\begin{aligned} EE &= \frac{discharge\ energy\ (Wh)}{charge\ energy\ (Wh)} \times 100\% = \frac{\int_{dis} I(t)V(t)dt}{\int_{ch} I(t)V(t)dt} \times 100\% \\ &= \frac{I_{dis} \bullet t_{dis}}{I_{ch} \bullet t_{ch}} \frac{\int_{dis} V(t)dt / t_{dis}}{\int_{ch} V(t)dt / t_{ch}} \times 100\% = CE \times VE \end{aligned} \tag{29}$$

Voltage efficiency (VE) is the average discharge voltage to the average charge voltage. **Figure 3a** shows the charging and discharging curves of VRFB in constant current mode, in which the current is maintained as constant value during charge-discharge cycle. While the current is constant during charge-discharge, the voltage is not constant but gradually changing in the whole cycle. Voltage efficiency represents a measure of electrical resistance loss and the polarization properties of battery. The polarization plot in **Figure 3c** coincides with the voltage efficiency trend in **Figure 3b**. Energy efficiency is the ratio of available energy to stored energy, which can be calculated as the product of voltage efficiency and current efficiency.

It is important to monitor the charging status of VRFB since especially overcharging the battery results in gas evolution side reactions, cell resistance increase, and capacity loss. Normally, VRFB is operated in charge range of 20–80%. The status of charge (SOC) is defined as the following using the concentrations of vanadium ions [8, 9]:

$$\mathrm{SOC} = \frac{C_{V^{2+}}}{C_{V^{2+}} + C_{V^{3+}}} = \frac{C_{VO_2^+}}{C_{VO_2^+} + C_{VO^{2+}}} \tag{30}$$

3. Electrode

The electrode provides the active sites for the redox reaction of redox couples dissolved in the electrolyte notwithstanding the electrode itself does not participate in the reaction. The electrode material influences the performance of VRFB diversely. The electrode should be electrochemically stable in the operating potential window of VRFB. The electrochemical activity of electrode affects the charge-discharge voltages and consequently the voltage efficiency during battery cycle operation. The electrode must have high electrical conductivity to increase the charge transfer speed. The charge transfer speed is related the ohmic losses, cell voltage, and energy efficiency. The vanadium can be dissolved in strong acidic aqueous solution; therefore the electrode should be chemically stable in strong acidic condition. The chemical stability of the electrode in acid electrolyte is related to the corrosion resistance when oxygen is generated at the positive electrode during overcharged and determines the lifetime of VRFB. The porosity of the electrode affects the pumping energy loss, which affects pressure drop across the stack and overall battery system efficiency [15, 16].

3.1 Reaction mechanism at carbon felt electrode

Various carbon materials including carbon felt, graphite felt, and carbon paper have been extensively studied as electrodes for VRFB. Especially, carbon felts are considered to be suitable for use as electrodes of VRFB because of their wide specific surface area, high electrical conductivity, high chemical stability, and wide operating potential window.

Sun and Skyllas-Kazacos reported that the C-OH functional group acts as an active site for oxidation of VO^{2+} and reduction of V^{3+} on the surface of the electrode [17, 18]. Oxidation and reduction mechanisms of the VO_2^+/VO^{2+} and V^{2+}/V^{3+}

redox couples at the electrode surface can be explained in three steps as shown in **Figure 4**. At first step of charge process, the vanadium ions are diffused from the bulk electrolytes to the vicinity of the electrodes and absorbed on the surface of the electrodes. The absorbed vanadium ions are connected to the electrode through the exchange with functional group hydrogen ions. In the second step, the electron and oxygen transfer reactions occur in the VO_2^+/VO^{2+} redox couple, and only the electron transfer reaction occurs in the V^{2+}/V^{3+} redox couple. At the positive electrode, an oxygen atom of C-O functional group moves to the VO^{2+}, and an electron of the VO^{2+} is transferred to the electrode following the C-O-V bond, and the oxidation number of vanadium ion increases from +4 to +5. At the negative electrode, an electron is transferred from the electrode to the V^{3+} along the C-O-V

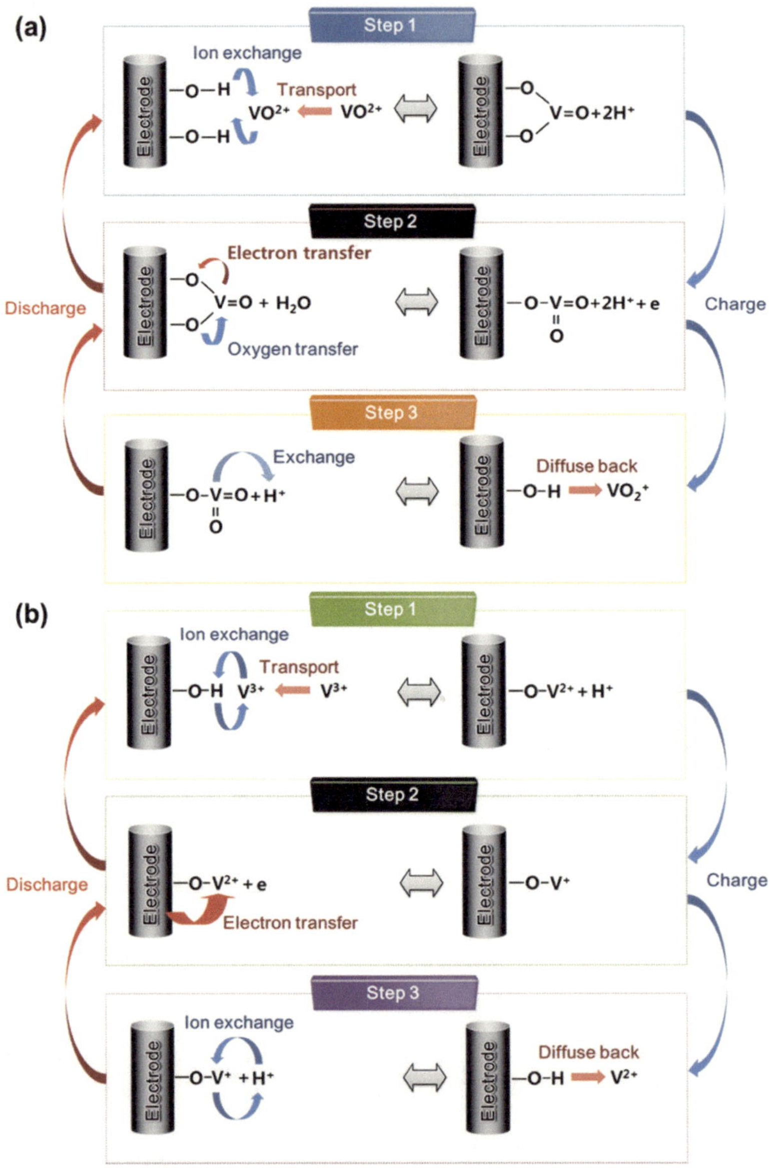

Figure 4.
Schematic illustration of the redox reaction mechanism for (a) VO_2^+/VO^{2+} redox couples in the catholyte and (b) V^{2+}/V^{3+} redox couple in the anolyte on the surface of the carbon felt electrode in VRFB. Reproduced with permission from [16]. Copyright 2015 by the Royal Society of Chemistry.

bond, and the oxidation number of vanadium ion is reduced from +3 to +2. In the third step, the ion exchange process between the V ion attached to the electrode surface and the H^+ ion in the electrolyte occurs, and the produced reactants (VO_2^+ and V^{2+}) diffuse back into the originated electrolytes, respectively.

To improve the electrochemical performance of VRFB, it is necessary to promote the reaction kinetics of vanadium ion redox couples. For this purpose, the electrode should have high electrical conductivity and the sufficient amount of oxygen and nitrogen functional groups at the surface.

3.2 Electrochemical characters

Cyclic voltammetry (CV) is used to monitor the reaction rates of redox couples and to evaluate the electrode performance of flow batteries. The CV curves in **Figure 5** show the electrode characteristics of the VRFB cell. The negative potential region of CV indicates the redox reaction of V^{2+}/V^{3+} ions, and the positive potential region implies the redox reaction of VO_2^+/VO^{2+} ions in electrolyte.

Figure 5a compares the electrode characteristics of the standard sulfuric acid electrolyte and the mixed acid electrolyte containing 6 M Cl^-. The peak current of the vanadium redox reaction is higher in the mixed electrolyte than in the standard sulfuric acid solution. This indicates that the reaction kinetics is improved due to the excellent fluidity of the electrolyte by adding sulfate chloride. The reaction voltage of the redox couples in the mixed solution increases slightly comparing to the sulfate solution, but there is no significant difference in the electrochemical reversibility between the sulfuric acid and the mixed electrolyte.

Figure 5b shows the reaction characteristics of carbon paper and catalytic behavior of biomass-derived activated carbon (AC) in the vanadium electrolyte. The V^{3+}/VO^{2+} redox couple peaks appear clearly in AC-coated carbon paper CV curve, and these multivalent peaks reveal the superior catalytic activity of AC coating.

Park et al. [21] investigated the change of VRFB performance according to the compression ratio of the carbon felt electrode and suggested the optimal compression ratio of the electrode. Oh et al. [22] conducted a numerical study of the VRFB model to investigate the effect of electrode compression on the charging and discharging behavior of VRFB. Yoon et al. [23] studied the flow distribution depending on local porosity of the electrode both numerically and experimentally.

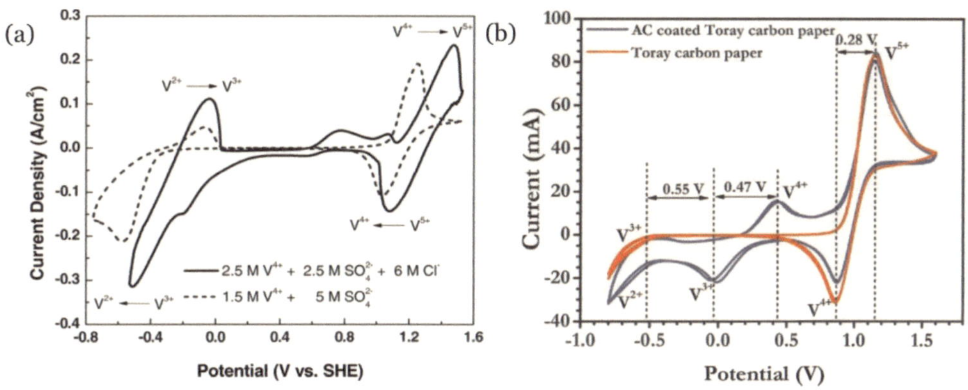

Figure 5.
(a) Cyclic voltammograms on a graphite felt electrode of a standard sulfate VRFB electrolyte (1.5 M V^{4+} and 5.0 M SO_4^{2-}) and a mixed electrolyte solution (2.5 M V^{4+}, 2.5 M SO_4^{2-}, and 6 M Cl^-) at a scan rate of 0.5 mV/s. Reproduced with permission with [19]. Copyright 2011 Wiley. (b) Cyclic voltammograms on Toray carbon sheets with and without mesoporous AC loading in the presence of 1.7 M $V^{3.5+}$ in 4 M H_2SO_4 solutions at a scan rate of 5 mV/s. Reproduced with permission from [20]. Copyright 2015 Elsevier.

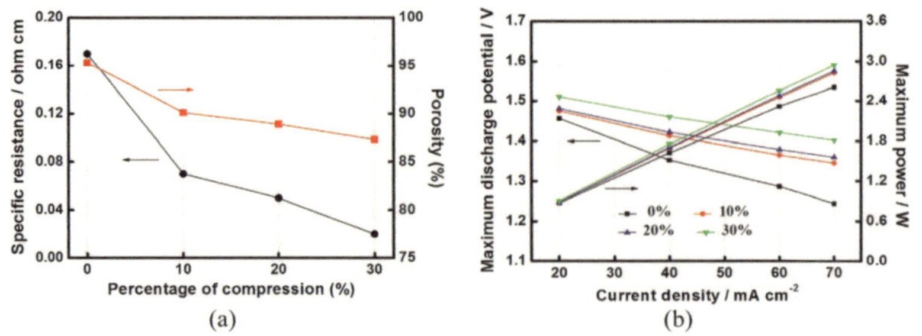

Figure 6.
(a) Specific resistance and porosity vs. percentage of compression for FA-30A carbon felt electrodes and (b) polarization curves of VRFB cells with electrodes of various levels of compression. Reproduced with permission from [21]. Copyright 2014 Elsevier.

As the percentage of electrode compression increases, the specific resistance and porosity of the electrode decrease as shown in **Figure 6a**. Compressed electrodes with reduced resistivity promote electron transfer, which increases the discharge time and maximum power of the VRFB cell and significantly increases VRFB performance efficiencies and discharge capacities, especially under high current density (**Figure 6b**). However, decreased porosity reduces the electrolyte flow passages through the electrode and increases pumping losses. The energy efficiency of the battery increases with increasing electrode compression ratio of up to 20%. When the carbon felt electrode is compressed more than 20%, the energy efficiency can be reduced due to the combined effect of deteriorated electrolyte transport and enhanced electron transfer. Overall, it can be concluded that the compression of the carbon felt electrode has a positive effect on cell performance, and the compression ratio optimization can generate significant improvement of VRFB performance without additional cost.

4. Electrolyte flow

The flow characteristics have a significant effect on the performance of redox flow battery. The flow distribution is related to the supply of reactant and participation of active species in redox reaction. The uniform flow distribution represents the uniform current density distribution. If the electrolyte flows nonuniformly, the reactants are not fully employed to the electrochemical reaction, which will lead to the degradation of the VRFB performance and durability.

Electrolyte flow rate is the speed of supplying reactants to the active site of electrode. If the flow rate is not enough, the capacity of the electrolytes is not fully utilized. If the flow rate is too high, the pumping loss increases, and the overall system efficiency is reduced accordingly. Therefore, optimizing flow rate is necessary in VRFB operation, and the importance increases significantly as storing capacity increases. The theoretical flow rate can be calculated as below [8]:

$$Q_{theo} = \frac{I}{n \times F \times C \times SOC_{min}} \tag{31}$$

where I is the current; n is the number of electrons transferred during the reaction, which is 1 for VFB; C is the total vanadium concentration for each reservoir (1.6 M); and SOC_{min} is the minimum state of charge, which is 20% normally.

The stoichiometric number, λ, is defined as the ratio of the actual flow rate to theoretical flow rate. **Figure 7** shows that as the stoichiometric number increases, the charge-discharge cycle time increases. The extension of the cycle time can be

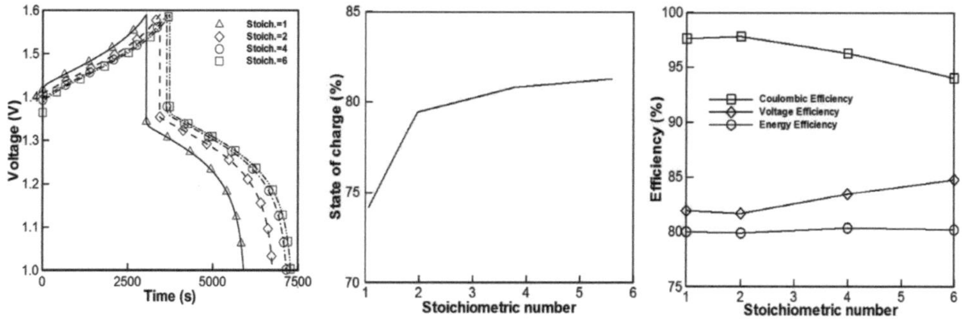

Figure 7.
Current density of 75 mA/cm² at various flow rates; (a) charge–discharge curve, (b) SOC, and (c) efficiencies as a function of stoichiometric number ($\lambda = Q_{real}/Q_{theo}$). Reproduced with permission from [8]. Copyright 2018 Elsevier.

explained as the increase in capacity of the VRFB, which means that the battery can store more energy. **Figure 7b** shows SOC increasing corresponding to the flow rate increase. On this basis, it is clear that a large mass flow rate can enhance the utilization of vanadium ions. This result explains the increase in the VFB capacity as the stoichiometric number increases. The variation of the efficiencies according to the flow rate is shown in **Figure 7c** and similar to the efficiency behavior according to the current density.

Flow patterns of RFB can be categorized into two types: "flow-through" type without flow field and "flow-by" type which has a flow field design on the bipolar plate. Leung et al. [25] explained that the structure in which the flow direction is parallel to the current direction is "flow-through" type and the structure in which the flow direction is perpendicular to the current direction is "flow-by" type. However, this definition does not match the concept we are dealing with here. In the scheme described here, the directions of electrolyte flow and electric current are perpendicular to each other in both "flow-through" and "flow-by" configurations. **Figure 8** shows the flow battery stack configuration and conceptual schematics of both flow designs. The classical "flow-through" type is the configuration in which the electrolyte flows through the porous carbon felt electrode. A "flow-by" type is the structure in which the electrolyte flows by the surface of an electrode following the flow field at the bipolar plate like a fuel cell. A "flow-by" type can choose relatively thinner carbon felt or carbon paper as an electrode material. Zawodzinski's group first reported better electrochemical performance and improved limiting current density and peak power density of VRFB with a "zero-gap" serpentine flow field design comparing to "flow-through" configuration [29]. This results from reduced ohmic loss and enhanced localized mass transfer due to thinner thickness and larger surface area-to-volume ratio of carbon paper used as electrode than those of carbon felt. Elgammal et al. [30] achieved normalized limiting current density of 2961 mA/cm^2 mol and peak power density of 2588 mW/cm^2 of VRFB with serpentine flow field. However, "flow-through" configuration distributes the electrolyte flow more uniformly and results in less pressure drops and pumping losses than "flow-by" configuration.

The electrolyte flow behavior is indicated schematically in **Figure 9**. The electrolyte is flowing mainly following channel over the electrode and partly penetrating into the porous electrode forced by pressure gradient. The flow velocity through the porous carbon media is lower than mean velocity of fully developed channel flow. The amount of the electrolyte penetrated into the porous electrode is associated with the stoichiometric availability of electrolyte reactants and the battery performance.

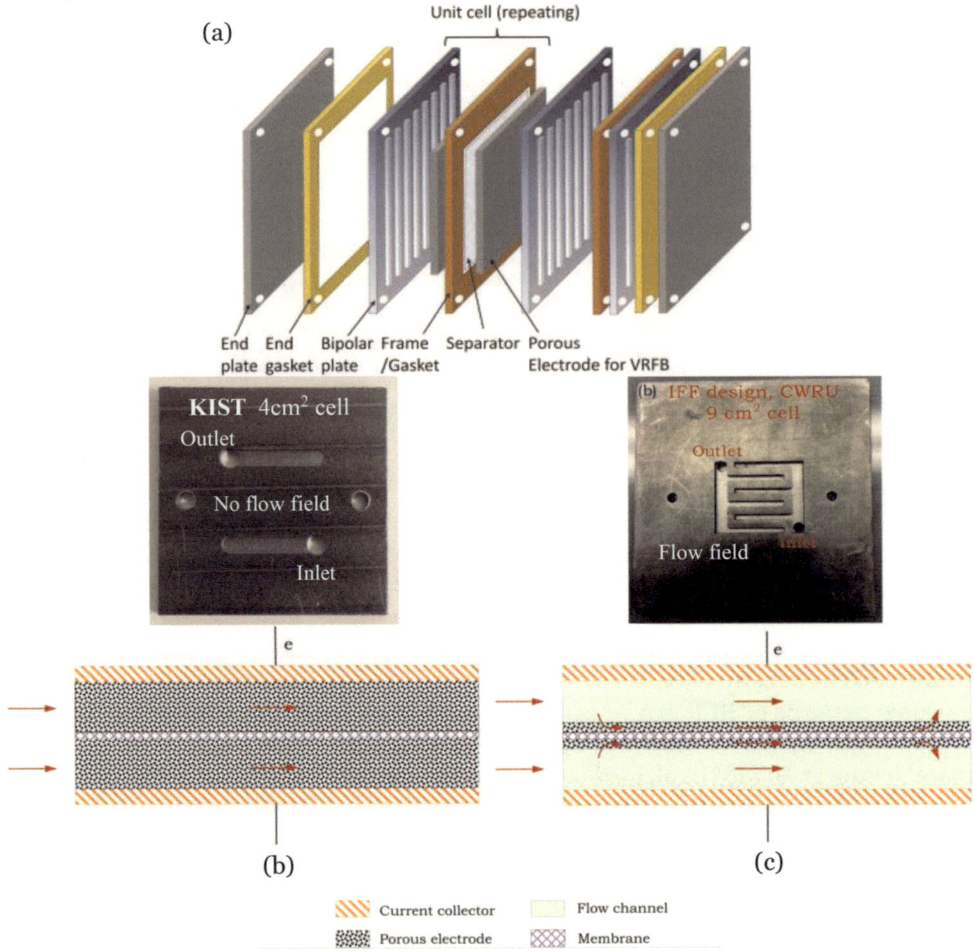

Figure 8.
(a) Schematic of flow battery stack configuration. Reproduced with permission from [31]. Copyright 2015 by Elsevier. (b) Bipolar plate and two-dimensional configuration of "flow-through" design and (c) "flow-by" design. Reproduced with permission from [26]. Copyright 2018 by the Royal Society of Chemistry.

Limiting current density is a key factor evaluating flow battery performance. High current density allows fast electrochemical reactions and reduces charging time. Newman et al. developed the limiting current density model as below [6]:

$$i_{lim} = 0.9783\frac{nFDc}{L}\int_{0}^{L}\left(\frac{u_f}{hDX}\right)^{\frac{1}{3}}dX \qquad (32)$$

where n is the number of electrons transferred during reactions, D is the diffusion coefficient, c is the bulk electrolyte concentration, L is the length of the flow channel, u_f is the averaged electrolyte flow velocity along the flow channel, and h is the distance between one electrode and one flat plate. Newman's model predicts the limiting current density of an electrolyte flowing between one flat plate and one electrode as shown in **Figure 9c** assuming no electrolyte penetration into the electrode surface.

The limiting current density dominated by the stoichiometric availability of reactant in the porous electrode as shown in **Figure 9d** is called "maximum current density" and can be expressed in Eq. (33) [26, 27]:

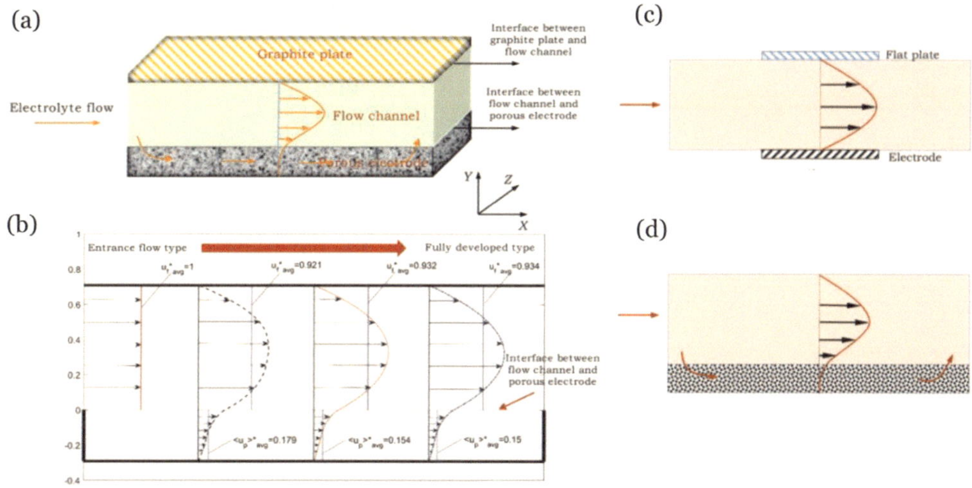

Figure 9.
(a) Diagram of electrolyte flow through a single flow channel and over the porous electrode in RFBs, (b) two-dimensional flow distributions in the flow channel-porous electrode layered system, and (c) the case of current density limited by the diffusion boundary layer formed between one flat plate and one electrode, which does not allow electrolyte reactant penetration. (d) the case of current density limited by the stoichiometric availability of the electrolyte reactants penetrate through the porous electrode from the flow channel. Reproduced with permission from [26]. Copyright 2018 by the Royal Society of Chemistry.

$$i_{max} = \frac{nFcQ_p}{A} \tag{33}$$

where Q_p is the volumetric flow of electrolyte reactant penetration through the interface between the flow channel and porous electrode and A is the cross-section area of porous electrode that is perpendicular to the current direction.

The entrance flow rate of "flow-by" type is higher than "flow-through" type. If entrance flow rate is increased, the penetrating electrolyte flow into the porous electrode is increased because the diffusion boundary layer is decreased, and the maximum current density is increased according to Eq. (33).

Zawodzinski et al. have shown how the discharge polarization curves of VRFB behave with the flow field and flow rate variations [28]. The flow-through type shows a limiting current density of 165 mA/cm ^{2} at an electrolyte circulation rate of 30 ml/min

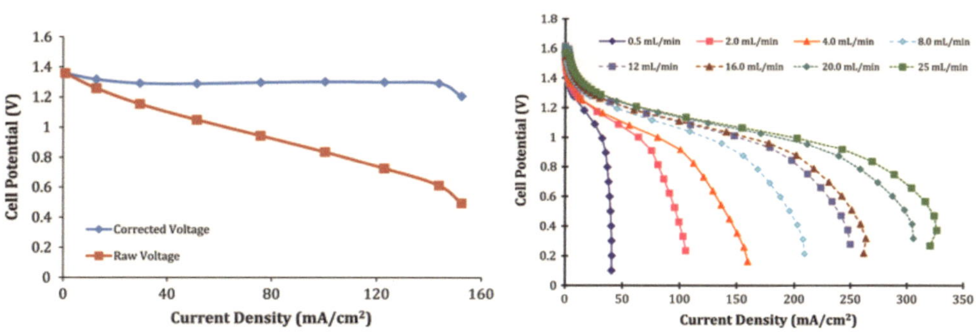

(a) (b)

Figure 10.
(a) Discharging polarization curve of the flow-through type VRFB (0.5 M V/2.0 M H_2SO_4 electrolyte with 30 ml/min) and (b) iR free discharge polarization curves illustrating the effect of the electrolyte flow rate on flow-by type VRFB (1.0 M V/5.0 M H_2SO_4 electrolyte). Reproduced with permission from [28]. Copyright 2011 by springer.

Flow rate (ml/min)	Theoretical limiting current density (mA/cm^2)	Observed limiting current density (mA/cm^2)	Percent of max current
0.5	161	40	25.2
2	643	105	16.3
4	1287	159	12.4
8	2573	209	8.12
12	3860	250	6.48
16	5147	261	5.07
20	6433	306	4.76
25	8042	321	3.99

Table 1.
Comparison of theoretical limiting current density and observed current density in flow-by configuration of VRFB at various electrolyte flow rates. Reproduced with permission from [28]. Copyright 2011 by springer.

(**Figure 10a**). **Figure 10b** shows that the limiting current density of the flow-by type increases from 40 to 321 mA/cm^2 as the flow rate increases from 0.5 to 25 ml/min. The values of the theoretical and observed limit current density according to the flow rate are summarized in **Table 1**. The theoretical limiting current density was calculated by converting the transfer rate of the electrolyte to the bipolar plate into the number of available electrons, assuming that all vanadium was converted in a single pass.

5. Cost analysis

Various batteries compete to become renewable energy storage devices in the power grid. One of the most important factors in practical implementation is the battery installation cost (capital cost). Noack et al. [32] conducted a techno-economic modeling analysis based on a 10 kW/120 kWh VRFB system. The costs and ratios of each component are summarized in **Table 2** and **Figure 11,** respectively. The largest portion of the VRFB cost is the stack, which accounts for 40% of

VRFB system parameter	Cost	VRFB stack component	Cost
Electrolyte	€ 41,000	Bipolar plate	€ 11,211
Tank	€ 9082	Felt electrode	€ 11,047
System assembling	€ 9000	Frame	€ 3066
Power electronics	€ 5000	Membrane	€ 6656
Fluid components	€ 3420	Gasket	€ 16,974
Control engineering	€ 9160	Assembling	€ 2782
VRFB stack	€ 52,646	End plate	€ 435
VRFB stack specific cost	€ 5265 /kW	Isolation plate	€ 217
Total system cost	€ 129,310	Current collector	€ 141
Total system specific cost	€ 1078 / kWh	Connection	€ 119

Table 2.
Cost analysis of 10 kW/120 kWh VRFB system. Reproduced with permission from [32]. Copyright 2016 by Noack J. et al.

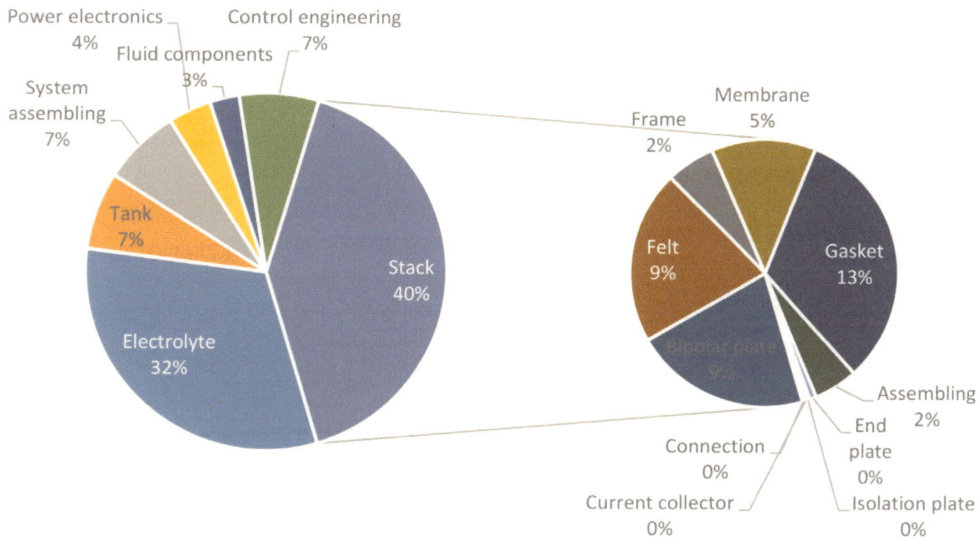

Figure 11.
10 kW/120 kWh VRFB system cost analysis. Reproduced with permission from [32]. Copyright 2016 by Noack J. et al.

the total cost. Electrolyte accounts for 32% of the total cost, which is the largest portion as a single component. In order to increase the energy content of the flow battery, the additional active material and the tank are required, so that the cost proportion of the electrolyte may increase depending on the storage capacity increase and the fluctuation of vanadium market price. In this analysis, the energy storage cost for VRFB system is presented at € 1078/kWh, which is expected to decrease with increasing production quantities.

6. Conclusions

Vanadium redox flow battery is one of the most promising devices for a large energy storage system to substitute the fossil fuel and nuclear energy with renewable energy. The VRFB is a complicated device that combines all the technologies of electrochemistry, mechanical engineering, polymer science, and materials science similar to the fuel cell. To optimize the flow battery design, it is necessary to understand the flow distribution, local current distribution, limits, and maximum current density. Understanding the shunt current and pressure distribution allows to design the flow battery stack with high power, large capacity, and high system efficiencies. Both experimental and modeling approaches are required to develop advanced vanadium redox flow battery stacks with high electrochemical performance.

Since Skyllas-Kazacos group at the University of New South Wales invented the VRFB in 1986, many researchers have conducted VRFB research. It is true that the VRFB are closer to commercialization than any other flow batteries. However still many of the reaction mechanisms and material characteristics must be further studied, and it is sure that the vanadium redox flow batteries are still very attractive research topics.

Acknowledgements

This research was supported by the basic research project of Korea Institute of Science and Technology (KIST) Europe, "Electrochemical energy transformation and energy storage".

Author details

Sangwon Kim[1,2]

1 Korea Institute of Science and Technology (KIST) Europe, Saarbrücken, Germany

2 Transfercenter Sustainable Electrochemistry, Saarland University, Saarbrücken, Germany

*Address all correspondence to: sangwon.kim@kist-europe.de

References

[1] Chen R, Kim S, Chang Z. Redox flow batteries: Fundamentals and applications. In: Khalid M, editor. Redox: Principles and Advance Applications. London: InTech; 2017. pp. 103-118. DOI: 10.5772/intechopen.68752

[2] Ye R, Henkensmeier D, Yoon SJ, Huang Z, Kim DK, Chang Z, et al. Redox flow batteries for energy storage: A technology review. ASME Journal of Electrochemical Energy Conversion and Storage. 2018;**15**:010801. DOI: 10.1115/1.4037248

[3] Choi C, Kim S, Kim R, Choi Y, Kim S, Jung H, et al. A review of vanadium electrolytes for vanadium redox flow batteries. Renewable and Sustainable Energy Reviews. 2017;**69**:263-274. DOI: 10.1016/j.rser.2016.11.188

[4] Alotto P, Guarnieri M, Moro F. Redox flow batteries for the storage of renewable energy: A review. Renewable and Sustainable Energy Reviews. 2014; **29**:325-335. DOI: 10.1016/j.rser.2013.08.001

[5] Bard AJ, Faulkner LR. Electrochemical Methods: Fundamentals and Applications. 2nd ed. Hoboken: Wiley; 2000. 832 p

[6] Newman J, Thomas-Alyea KE. Electrochemical Systems. 3rd ed. Hoboken: Wiley; 2004. 647 p

[7] Zhang H, Li X, Zhang J. Redox Flow Batteries: Fundamentals and Applications. Boca Raton: CRC Press; 2018. 432 p

[8] Kim D, Yoon S, Lee J, Kim S. Parametric study and flow rate optimization of all-vanadium redox flow batteries. Applied Energy. 2018; **228**:891-901. DOI: 10.1016/j.apenergy.2018.06.094

[9] Knehr KW, Kumbur EC. Open circuit voltage of vanadium redox flow batteries: Discrepancy between models and experiments. Electrochemistry Communications. 2011;**13**:342-345. DOI: 10.1016/j.elecom.2011.01.020

[10] Hwang G, Kim S, In D, Lee D, Ryu C. Application of the commercial ion exchange membranes in the all-vanadium redox flow battery. Journal of Industrial and Engineering Chemistry. 2018;**60**:360-365. DOI: 10.1016/j.jiec.2017.11.023

[11] Jung M, Lee W, Krishan NN, Kim S, Gupta G, Komsyska L, et al. Porous-Nafion/PBI composite membranes and Nafion/PBI blend membranes for vanadium redox flow batteries. Applied Surface Science. 2018;**450**:301-311. DOI: 10.1016/j.apsusc.2018.04.198

[12] Chen C, Henkensmeier D, Kim S, Yoon SJ, Zinkevich T, Indris S. Improved all-vanadium redox flow batteries using catholyte additive and a cross-linked methylated polybenzimidazole membrane. ACS Applied Energy Materials. 2018;**1**: 6047-6055. DOI: 10.1021/acsaem.8b01116

[13] Strużyńska-Pirona I, Jung M, Maljuschc A, Conradic O, Kim S, Jang J, et al. Imidazole based ionenes, their blends with PBI-OO and applicability as membrane in a vanadium redox flow battery. European Polymer Journal. 2017;**96**:383-392. DOI: 10.1016/j.eurpolymj.2017.09.031

[14] Lee Y, Kim S, Hempelmann R, Jang JH, Kim HJ, Han J, et al. Nafion membranes with a sulfonated organic additive for the use in vanadium redox flow batteries. Journal of Applied Polymer Science. 2019;**136**:47547. DOI: 10.1002/app.47547

[15] Ulaganathana M, Aravindan V, Yan Q, Madhavi S, Skyllas-Kazacos M, Lim TM. Recent advancements in all-vanadium redox flow batteries. Advanced Materials Interfaces. 2015;**3**: 1500309. DOI: 10.1002/admi.201500309

[16] Kim KJ, Park MS, Kim YJ, Kim JH, Dou SX, Skyllas-Kazacos M. A technology review of electrodes and reaction mechanisms in vanadium redox flow batteries. Journal of Materials Chemistry A. 2015;**3**:16913-16933. DOI: 10.1039/C5TA02613J

[17] Sum E, Skyllas-Kazacos M. A study of the V(II)/V(III) redox couple for redox flow cell applications. Journal of Power Sources. 1985;**15**:179-190. DOI: 10.1016/0378-7753(85)80071-9

[18] Sum E, Rychcik M, Skyllas-Kazacos M. Investigation of the V(V)/V(IV) system for use in the positive half-cell of a redox battery. Jounral of Power Sources. 1985;**16**:85-95. DOI: 10.1016/0378-7753(85)80082-3

[19] Li L, Kim S, Wang W, Vijayakumar M, Nie Z, Chen B, et al. A stable vanadium redox-flow battery with high energy density for large-scale energy storage. Applied Energy Materials. 2011; **1**:394-400. DOI: 10.1002/aenm.201100008

[20] Ulaganathana M, Jain A, Aravindan V, Jayaraman S, Ling WC, Lim TM, et al. Bio-mass derived mesoporous carbon as superior electrode in all vanadium redox flow battery with multicouple reactions. Jounral of Power Sources. 2015;**15**:846-850. DOI: 10.1016/j.jpowsour.2014.10.176

[21] Park SK, Shim J, Yang JH, Jin CS, Lee BS, Lee YS, et al. The influence of compressed carbon felt electrodes on the performance of a vanadium redox flow battery. Electrochimica Acta. 2014; **116**:447-452. DOI: 10.1016/j.electacta.2013.11.073

[22] Oh K, Won S, Ju H. Numerical study of the effects of carbon felt electrode compression in all-vanadium redox flow batteries. Electrochimica Acta. 2015;**181**: 13-23. DOI: 10.1016/j.electacta.2015.02.212

[23] Yoon SJ, Kim S, Kim DK. Optimization of local porosity in the electrode as an advanced channel for all-vanadium redox flow battery. Energy. 2019;**172**:26-35. DOI: 10.1016/j.energy.2019.01.101

[24] Weber AZ, Mench MM, Meyers JP, Ross PN, Gostick JT, Liu Q. Redox flow batteries: A review. Journal of Applied Electrochemistry. 2011;**41**:1137-1164. DOI: 10.1007/s10800-011-0352-6

[25] Leung P, Li X, Leon CP, Berlouis L, Low CTJ, Walsh FC. Progress in redox flow batteries, remaining challenges and their applications in energy storage. RSC Advances. 2012;**2**:10125-10156. DOI: 10.1039/C2RA21342G

[26] Ke X, Prahl JM, Alexander JID, Wainright JS, Zawodzinski TA, Savinell R. Rechargeable redox flow batteries: Flow fields, stacks and design considerations. Chemical Society Reviews. 2018;**47**:8721-8743. DOI: 10.1039/C8CS00072G

[27] Ke X, Alexander JID, Prahl JM, Savinell RF. Flow distribution and maximum current density studies in redox flow batteries with a single passage of the serpentine flow channel. Journal of Power Sources. 2014;**270**: 646-657. DOI: 10.1016/j.jpowsour.2014.07.155

[28] Aaron A, Tang Z, Papandrew AB, Zawodzinski TA. Polarization curve analysis of all-vanadium redox flow batteries. Journal of Applied Electrochemistry. 2011;**41**:1175-1182. DOI: 10.1007/s10800-011-0335-7

[29] Aaron DS, Liu Q, Tang Z, Grim GM, Papandrew AB, Turhan A, et al. Dramatic performance gains in

vanadium redox flow batteries through modified cell architecture. Journal of Power Sources. 2012;**206**:450-453. DOI: 10.1016/j.jpowsour.2011.12.026

[30] Elgammal RA, Tang Z, Sun CN, Lawton J, Zawodzinski TA. Species uptake and mass transport in membranes for vanadium redox flow batteries. Electrochimica Acta. 2017;**237**: 1-11. DOI: 10.1016/j.electacta.2017.03.131

[31] Ha S, Gallagher KG. Estimating the system price of redox flow batteries for grid storage. Journal of Power Sources. 2015;**296**:122-132. DOI: 10.1016/j.jpowsour.2015.07.004

[32] Noack J, Wietschel L, Roznyatovskaya N, Pinkwart K, Tübke J. Techno-economic modeling and analysis of redox flow battery systems. Energies. 2016;**9**:627. DOI: 10.3390/en9080627

Chapter 8

Hydrogen Energy Storage

Dallia Mahmoud Morsi Ali

Abstract

The dominating trend of variable renewable energy sources (RES) continues to underpin the early retirement of baseload power generating sources such as coal, nuclear, and natural gas steam generators; however, the need to maintain system reliability remains the challenge. Implementing energy storage with conventional power plants provides a method for load leveling, peak shaving, and time shifting allowing power quality improvement and reduction in grid energy management issues, implementing energy storage with RES smooth their intermittency, by storing the surplus in their generation for later use during their shortfall, thus enabling their high penetration into the electricity grid. Energy storage technologies (EST) can be classified according to many criteria like their application (permanent or portable), capacity, storage duration (short or long), and size (weight and volume). EST suited for short duration storage and low-to-medium power outputs are seen performing better in improving power quality, while those providing medium-to-high power outputs with long durations are seen better suited for energy management of electrical networks. With the growing deployment of renewable energy systems, EST must be utilized to allow the grid to absorb the increased integration of RES generation. The recent advances in hydrogen energy storage technologies (HEST) have unlocked their potential for use with constrained renewable generation. HEST combines hydrogen production, storage, and end use technologies with the renewable generation either in a directly connected configuration or in an indirectly connected configuration via the existing power network. This chapter introduces the hydrogen energy storage technology and its implementation in conjunction with renewable energy sources. The efficiency of renewable hydrogen energy storage systems (RHESS) will be explored with a techno-economic assessment. A levelized cost (LC) model that identifies the financial competitiveness of HEST in different application scenarios is given, where five scenarios are investigated to demonstrate the most financially competitive configuration. To address the absence of a commercial software tool that can quickly size an energy system incorporating HEST while using limited data, a deterministic modeling approach that enables a quick initial sizing of hybrid renewable hydrogen energy systems (HRHES) is given in this chapter. This modeling approach can achieve the initial sizing of a HRHES using only two input data, namely the available renewable energy resource and the load profile. A modeling of the effect of the electrolyzer thermal transients at start-up, when operated in conjunction with an intermittent renewable generation, on the quantity of hydrogen produced is also given in this chapter.

Keywords: hydrogen energy storage technology, renewable hydrogen energy storage systems, levelized cost modeling, sizing hybrid renewable hydrogen energy system for a specified demand and renewable resource, modeling the effect of the electrolyzer thermal transients at start-up when powered by renewables on the quantity of hydrogen produced

1. Introduction

The green-house gas emissions associated with conventional electricity generation will lead to an increase in the average global temperature over the upcoming years, which in turn will lead to raised sea levels and more frequent extreme weather conditions and droughts. To mitigate such global climate changes, the world needs an energy transition that allows a cleaner and more sustainable energy supply.

The decarbonization of the world's energy system has started in 2015 after the signature of the legally binding agreement by 195 countries to keep the global warming well below 2°C [1]. Since then, significant amounts of RES have been installed and integrated into the grid while securing the supply and the system resilience. Expanding the utilization of RES into the electricity grid however requires large-scale electricity storage to cover for their energy intermittency. Even a small-scale grid that handles only 10–30 gigawatts could not rely entirely on intermittent RES without having a gigawatt-scale storage that can work for many hours; so, for example, securing 3 GW for 2 days requires a 144 gigawatt-hours storage. There are many storage options; some of these are the flow batteries, which store the energy directly in the electrolyte, but are still in an early stage of deployment; sodium-sulfur batteries, with higher energy density than Li-ion ones; however, their hot liquid metal electrolyte is inconvenient; supercapacitors, which cannot provide electricity over a long enough time; and compressed air and flywheels have made it only to small and midsize installations due to location restrictions. Hydrogen energy storage however offers a clean, sustainable, and flexible storage option that can be scaled up to enable large-scale energy storage over long periods of time with no restrictions on location, and therefore it has the potential to enable the energy transition. While the transition toward using more variable RES into the power grid will unbalance the supply and demand, using the excess in the RES power supply in electrolysis to produce hydrogen and store it for future use during RES supply deficit can help balancing the grid. The stored hydrogen can also be used in other sectors like transport, industry, residential heat, etc. Implementing hydrogen energy storage with renewables therefore have the potential to improve the economic efficiency of renewable investments, enhance the security of power supply, and serve as a carbon-free seasonal storage supplying energy when the RES production is low or the energy demand is high.

This chapter explores the context of hydrogen as an energy vector and the role of hydrogen energy storage in the world energy transition. An exploration into the hydrogen technology techno-economic potential, its applications, achievements, and challenges to its deployment as well as recommendations for accelerating its deployment are covered in this chapter. This chapter also includes a model that has been developed to enable the quick sizing of a hybrid renewable hydrogen energy system (HRHES) that integrates solar and wind renewable resources combined with hydrogen energy storage to meet a specific electrical load. The effect of the electrolyzer thermal transients at start-up, when operated in conjunction with the intermittent renewable generation, on the quantity of hydrogen produced is included in the developed model. Implementing the developed model as a tool for identifying the performance issues within installed hydrogen systems during their operation is also given in this chapter. Two case studies are provided to verify the developed model, and to validate that thermally compensated electrolyzer models are essential for designing new hydrogen installations as well as for monitoring the performance issues within running installations.

2. Preface: role of energy storage in the world energy transition

Twenty-four percent of the global green-house gas emissions are produced from the electrical power generation sector [2]. Implementing RES into the power generation sector can play a vital role in reducing this emission percentage and accordingly address the climate change with its associated political, economic, and environmental pressures [3]. To enable the high penetration of fluctuating RES into the electrical power network, the network should be able to absorb and store the excess in the power fed so that they remain stable and should utilize the RES generation in the most possible effective manner. Delivering the RES generation to the load when needed and to the storage when the generation exceeds consumption allows the absorption of the excess in RES generation and thus reduces the need for the grid weak interconnections upgrading and remove a considerable amount of constraint issues. Additionally, implementing energy storage reduces the spinning reserve requirements and thus allows spinning reserve operational costs to be diverted to the EST operational costs [4].

The flexibility that the storage brings to the grid reduces the electrical supply and demand imbalance associated with increased RES integration, and thus facilitates energy transition. While the inclusion of energy storage brings some additional capital and operational costs together with energy conversion loss efficiencies [5] and although there is limited experience with implementing large-scale energy storage technologies (except for pumped hydro), the implementation of EST is still considered vital for enabling the projected increase of RES into the power network as a step toward achieving the energy transition [6].

2.1 The need for energy storage in modern power systems

Energy storage increases the power grid capacity in accommodating the increasing fluctuations in supply and demand, and thus it plays a crucial role in supporting the wider integration of distributed RES in modern electrical networks [7]. Integrating EST into electrical networks allows more flexibility in accommodating the increased amounts of RES [8].

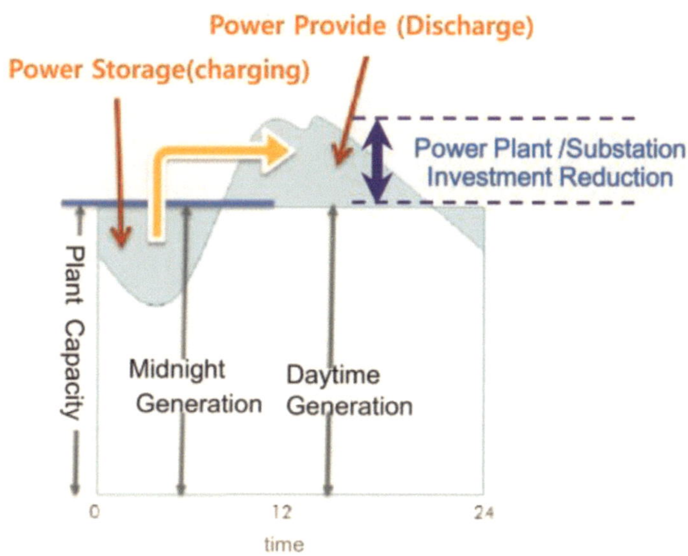

Figure 1.
Energy storage for Load Leveling and Peak Shaving.

In power systems, energy storage provides a method for "Load Leveling" by storing the power during periods of light loading and delivering it during periods of high demand, thus avoiding the high costs during peak demand and postponing investments in grid upgrades or in building new generating capacity. It also provides a method for "peak shaving," which works like load leveling but aims to reduce the peak demand. Energy storage can also be used for "time shifting" by storing the energy during low price times, and discharging it during high price times. All these actions allow reduction in the grid's energy management issues and improve the power quality. **Figure 1** depicts how energy storage allows load leveling and peak shaving with conventional power plants, and **Figure 2** depicts how implementing bulk energy storage with intermittent RES facilitates their high

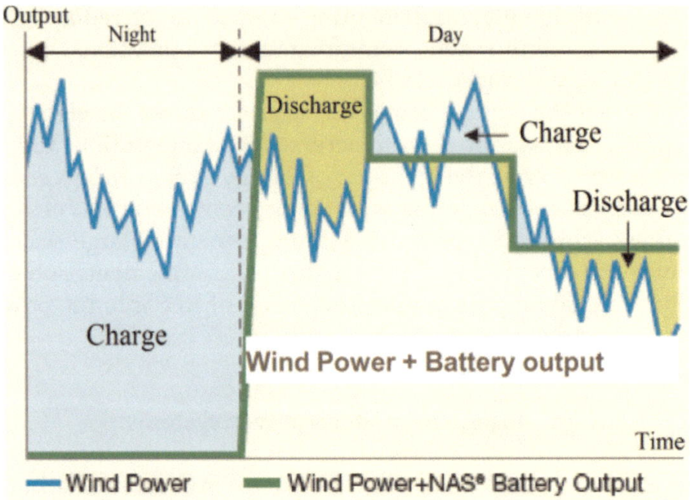

Figure 2.
Energy storage with RES to enable their high grid penetration.

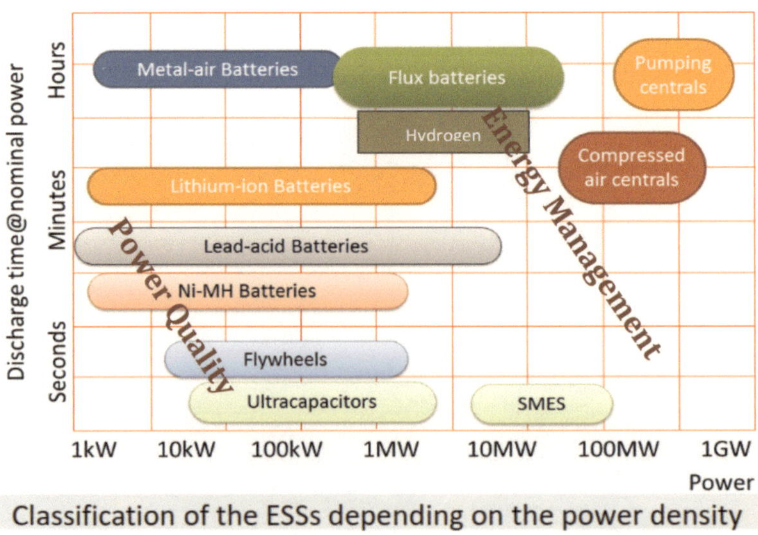

Figure 3.
*Energy storage technology overview, Scottish Government Report [10]. *SMES = superconducting magnetic energy storage.*

penetration into the electricity grid through storing the surplus in their energy to be used later to match the fluctuating load demand curve.

Energy storage is crucial for many applications and while implementing them in large size at the supply side can assist in the network bulk energy management, implementing them distributed near the consumer can assist in reducing power quality issues. EST energy storage duration ranges from few seconds of operation to several hours [9]. **Figure 3** shows a summary of the different storage technologies with the applications in which they are best suited as conducted by the AEA technologies for the Scottish Government [10].

It can be seen from **Figure 3** that energy storage technologies that are best suited for short duration storage and low to medium power outputs are typically seen as performing better in improving the power quality, while EST that provide medium to high power outputs with long durations are better suited for the energy management of electrical networks.

3. Context for hydrogen as an energy vector

Traditionally, the electrical network infrastructure has been designed to deliver electricity from several large-scale centralized fossil-fuelled electrical power stations to several domestic, industrial, and commercial consumers through a transmission network that is suited for power flow in one direction from generation to load. The hydrocarbon-fuelled and nuclear power stations are conventionally load following and can adjust their electrical generation output to follow demand, and thus the electrical grid maintains its equilibrium. When the network experiences any imbalance condition, the network operators implement control mechanisms to return the network to its balanced state. So, when generation cannot meet demand, the spinning reserve is connected to the electrical network and is loaded up according to demand. However, spinning reserve is a highly costly, carbon intensive, and inefficient method for maintaining the network stability as it requires the generating stations to remain running and consuming fuel to be brought online very quickly when needed to maintain the supply and demand balance.

The admission of the RES into the grid provides genuinely green energy at the points of entry; however, it creates a huge operational problem in managing the central generators to cover any transient variations in the renewable power input and consumer demand, thus hindering the renewable potential. The power network operators will have to manage the imbalance by either increasing the operation of the spinning reserve leading to increased costs [11], or by implementing other ways different from what they use with traditional power stations [12]. So, owners of RES connected to an increasingly congested electrical transmission and distribution network will have to incur financial penalties when they produce power at times of low demand and the network operators will have to pay compensations to RES owners when they introduce constraints [13]. An example of this is what happened at Scotland in April 2014, when strong winds made the Scottish grid not able to absorb all the wind power generated and had to constrain it off the grid while paying compensations to the owners of wind generators. Approximately £890,000 was paid over few hours to six wind farms.

Given that the global renewable energy contribution (excluding bio-fuels) is predicted to increase by over 50% between 2010 and 2035 [14], the electrical power networks therefore need to have the capacity to store the excess power fed into the grid from fluctuating and intermittent power sources [15]. The electrical power networks also need to continually achieve equilibrium under the increasing demand conditions while reducing their dependence on the expensive and inefficient spinning reserve, and thus it is crucial to increase their capacity through the implementation of energy storage technologies [16]. Therefore, it can be concluded that implementing EST is essential in modern power grids.

Hydrogen energy storage technologies are slowly but surely unlocking the potential of RES. Integrating HEST into the power networks not only allow the absorption of excess energy from fluctuating RES, but also allow the supply of supplementary energy needed when the RES production is insufficient to meet demand. Thus, HEST enable balancing the supply and demand while allowing the increased implementation of variable output RES.

4. The hydrogen energy storage technology

Chemical energy storage in the form of hydrogen (gas or liquid) has the potential to store energy over long periods of time and can be scaled up with no restrictions on its location. Hydrogen can be used as an energy carrier, stored and delivered to where it is needed. The storage mechanism does not have high rate of self-discharge or degradation in performance. The basic elements of a hydrogen energy storage system (HESS) can be recognized in **Figure 4**. The electrolyzer (hydrogen generator) is used to convert the electrical energy from an energy source (typically renewable) into hydrogen for storage. The hydrogen storage system can store the hydrogen in several forms (pressurized gas, metal hydride, or liquid Dewar tank). A hydrogen energy conversion system then converts the stored chemical energy in the hydrogen back to electrical energy while giving off water and heat as by-products with no carbon emissions. The hydrogen energy conversion system that is commonly used is the fuel cell, given that its typical average electrical conversion efficiency, as recorded for installed projects, ranges between 40 and

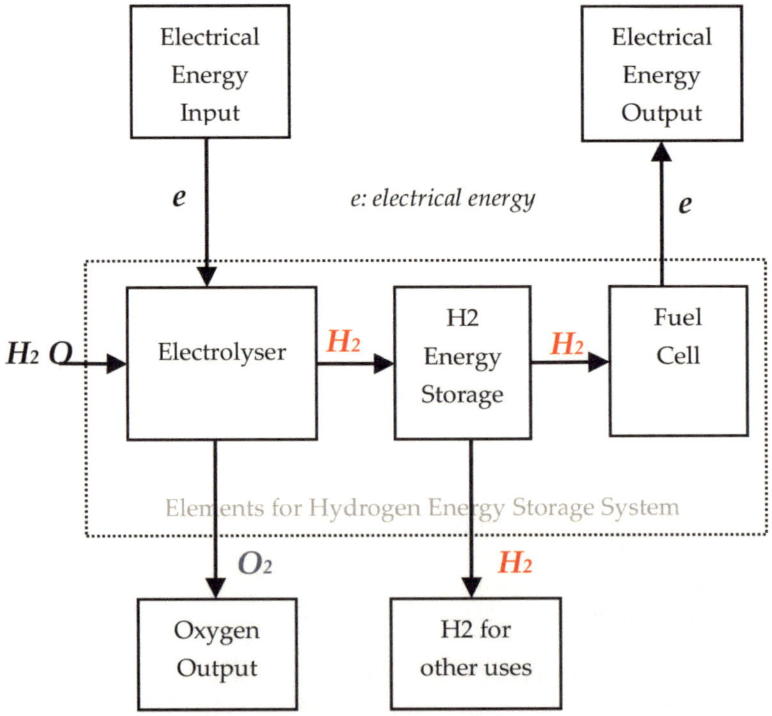

Figure 4.
Basic elements of a hydrogen energy storage system (HESS).

50% compared with a maximum of 37% for a small combustion engine [17]. Alternatively, the stored hydrogen can be used for other end uses and thus hydrogen and oxygen gases are sold as commodities.

4.1 Challenges to the hydrogen energy storage deployment

A key barrier to realize the potential of hydrogen energy storage systems is the limitation in the available modeling software and tools [18]. Another challenge is the ability to quantify the energy capacity and economic viability of the hydrogen energy storage technology (HEST) when integrated into the electrical power grid to enable the projected increase of renewables. Addressing these challenges is the main key for accelerating the wide deployment of the hydrogen technology.

4.2 Potential of the hydrogen energy storage technology

4.2.1 Role of hydrogen energy storage in allowing increased integration of renewable energy generation in constrained power networks

Hydrogen, as a form of energy storage, can deliver a fuel for making power or heat or for fueling a car while absorbing the intermittent power inputs from RES. Hydrogen production systems (electrolyzers) can be operated as deferrable and controllable loads within a smart grid infrastructure to allow the absorption of increased renewable energy generation in constrained power networks. The stored hydrogen can be used later in generating electricity when needed, or it can be used in other energy intensive sectors such as the gas grid, transport as a fuel, and industrial processes. Hydrogen storage is not geographically restricted and offers the potential to shift constrained renewable generation into other energy intensive sectors.

Large industrial and commercial consumers can play a vital role in balancing the grid through the intelligent use of their electrical loads while implementing hydrogen production and storage technologies. One example which demonstrates that hydrogen technology can be used for balancing the grid is what happens in "Tessenderlo Group," a company which utilizes both oxygen and hydrogen gases in its chemical processing activities [19]. Tessenderlo utilizes one of the world's largest scale hydrogen production systems, in the order of multiple Mega Watts (MW) scale [20, 21], and is charged at a lower cost/kWh in return to allow the local distribution network operator (DNO) to adjust its electrolytic hydrogen and oxygen production in maintaining the electrical network in the balanced state [22]. The DNO makes these adjustments in accordance with the demands on the electrical network using demand side management (DSM) techniques. The hydrogen production is reduced when the electrical network is experiencing a period of high demand and low energy production, and increased when the generation in electrical network exceeds consumption. Such a trading arrangement with a preferential tariff minimizes the need for the local network operators to waste money on highly inefficient spinning reserves. So, while utilizing the electrolyzer in maintaining the grid balance both hydrogen and oxygen gases are produced to be used in the chemical processes.

4.2.2 Position of the hydrogen energy storage technology

Hydrogen is in a strong position to be applied widely as an energy storage vector for balancing the grid while increasing the RES integration. Over the last decade, several renewable hydrogen concepts have been investigated [23] and several installations have been implemented to demonstrate the role of energy storage in the form of hydrogen in balancing the supply and demand in constrained grids. Many of these installations were based around small-scale RES of only a few tens of kilowatts, with exceptions to the hydrogen mini grid system (HMGS) in Rotherham, the Yorkshire [24, 25], the Utsira (Norway) energy system [26], and the Hydrogen Office [27], where large-scale RES have been utilized. All these systems have utilized commercially available alkaline electrolyzers with rated hydrogen production capacity in the range of 0.2–10 Nm^3/h and operating pressures in the range of 7–20 bar, except the Hydrogen Office electrolyzer of 3.5 Nm^3/h at 55 bar.

Hydrogen storage technologies can be divided into physical storage, where hydrogen molecules are stored (this includes pure hydrogen storage via compression and liquefaction), and chemical storage, where hydrides are stored. While chemical storage could offer high storage performance due to the strong binding of hydrogen and the high storage densities, the regeneration of storage material remains an issue with a large number of chemical storage systems still under investigation.

Demonstration projects have showed that hydrogen has a flexibility with RES, which is not available in other energy storage technologies. It has been found out that energy storage employing hydrogen technologies is best suited with renewable energy sources through the absorption of their surplus generation via electrolysis and storing it in the form of compressed hydrogen gas for later re-use in many applications such as the following:

- Controllable generation reserve via fuel cells and/or gas turbines and/or internal combustion engines (ICE)
- Fuel for transport applications
- A means to transfer the renewable energy into the gas grid
- A chemical process gas for other end uses in food, fertilizers, etc.

As the governments around the world are strategically moving toward a low carbon economy, hydrogen storage will undoubtedly play an important role in making use of the grid constrained “green” energy within a rapidly growing market [28].

4.2.3 Opportunities of the hydrogen energy storage technology

Hydrogen can be stored for long periods of time without degradation. Hydrogen can be stored in a gaseous or liquid form, or in some instances adsorbed onto a solid form in the case of metal hydride storage technology. Hydrogen is mixable with other gases making it suitable for mixing into the existing natural gas grid in a process known as sector shifting [29]. Additionally, hydrogen can be used as reactive agent in the chemical transformations of synthetic natural gas and fuels.

Hydrogen is seen by many of the energy industry experts as a mean of storing the surplus renewable energy from sources such as wind, solar, wave, and tide [30, 31] for later use. It is also seen to have a market potential for vehicle fueling in both urban and remote rural areas [32, 33].

4.2.4 Additional benefits of the hydrogen energy storage technology

It can be noticed from **Figure 3**, which has overviewed the different storage technologies together with the applications in which they are best suited as conducted by the AEA technologies for the Scottish government, that hydrogen has

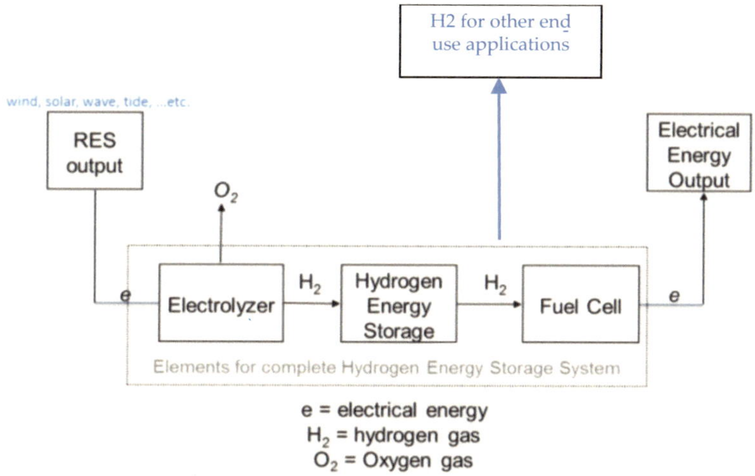

Figure 5.
Implementing hydrogen energy storage with RES.

a role to play in durations between several minutes to hours and is best suited for applications larger than 100 kW, and thus can be identified as appropriate for the Energy Management of Electricity Networks.

HES, when compared to the other ESTs, is seen to be suitable for use with RES [34]. In summary, the stored hydrogen produced during the RES excess generation periods can be:

- injected into the gas grids (since it is mixable with other gases);
- used to generate electricity and heat via a fuel cell;
- used to power a fuel cell (FC) or a combustion engine' vehicle; or
- used in many industrial processes (like fertilizer production)

Figure 5 overviews the implementation of hydrogen energy storage with RES.

4.2.5 Limitations toward the adoption of the hydrogen energy storage technology

Despite of the benefits and potential that HEST presents, the high capital cost and the low turn-around efficiency (i.e., electricity to hydrogen stored then back to electricity) are two noteworthy limitations [35]. Significant efforts are being made by industry to address cost and efficiency concerns. Additionally, many countries have started the process of publishing draft guidelines for the use of hydrogen energy storage technologies [36].

To contribute to the effective and wider implementation of the hydrogen technology, especially where HESS are operated in combination with variable RES, appropriate financial mechanisms and effective modeling techniques are developed in this chapter.

5. Techno-economic assessment of hydrogen energy storage

Energy storage technologies are generally compared in terms of their lifetime, efficiency, energy density, power density, and technological-maturity [37].

They are also often compared based on application-specific benefits and specific characteristics of interest [38, 39]; however, such comparisons did not take into consideration their financial competitiveness.

Financial competitiveness of EST is to define the price of stored energy per kWh over the lifetime of the energy storage system. The Electric Power Research Institute (EPRI) has developed and documented a method that analyses the costs associated with grid connected energy storage applications [40]. The EPRI utilizes a levelized cost model (LCM) approach to perform the cost benefit analysis for energy storage technologies. The LC, in its basic form, is calculated by dividing the annual expenditures by the annual income and correcting for inflationary effects. The LC reflects the capital and operational expenditures including the upfront capital costs, the fuel expenses, the operating and maintenance (O&M) charges, the financing costs, etc. levelized cost analysis is often used in regulatory review and longer-term resource planning [41]. Levelized costs can be done using limited input data, thus useful for evaluating technologies with limited operating experience or available data.

The HEST has not been included in the EPRI analysis [42] or in other cost analysis modeling techniques available in literature. HEST appears to be commonly excluded from the cost comparative studies due to its high capital cost and low turnaround efficiency compared to other bulk ESTs. Even the studies that examined HEST have not included its additional revenue streams. So, a study that has been completed by the National Renewable Energy Laboratories (NREL) has introduced the value of grid connected stored hydrogen for transport applications [43], but did not consider the potential value of the oxygen gas as a by-product from electrolysis. Since the by-product oxygen would either have a positive or a negative effect on the cost competitiveness of HEST, therefore, a new LCM is developed in this section to explore this.

Figure 6 illustrates the economic revenue streams of both the conventional and hydrogen energy storage technologies. While conventional energy storage systems allow for energy to be stored and released in the form of electrical energy, hydrogen as an energy storage mechanism allows surplus RES electrical energy to be stored and released as electricity in addition to hydrogen and oxygen gases that could be sold as commodities offering greater financial competitiveness. Since this could offset the HEST low energy efficiency and high capital costs, it is considered in the developed HEST financial competitiveness model.

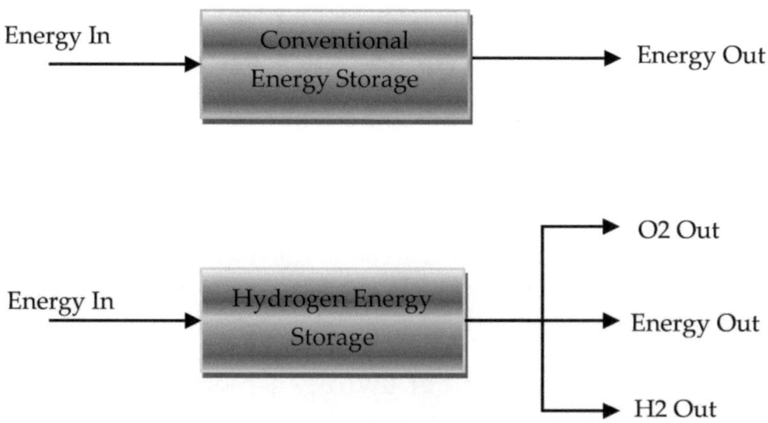

Figure 6.
HEST possible economic revenue streams compared to conventional EST.

5.1 The proposed levelized cost model (LCM) for HEST

The costs of energy storage are considered using a levelized cost of ownership approach. Typically, the capital costs of an energy storage facility are expressed as £/kW installed, where it includes all expenses involved in the purchase and installation of facility. The £/kW capital expenditure (CapEx) multiplied by the size of the facility produces the total cost of the project. In the proposed model, all the costs related to an energy storage facility are expressed as (i) total £/kW of usable discharge capacity (in kW) and (ii) total £/kWh of usable energy storage capacity. EST with deeper Depth of Discharge (i.e., the ability of an ESS to release its stored energy) and higher turn-around efficiency (i.e., ratio between input energy and output energy) will have a lower unit cost of usable power and energy [44].

Using the levelized cost approach, the levelized storage cost (LSC) of energy storage technology can be expressed as shown in Eq. (1) [41]:

$$LSC = \frac{\sum_{t=1}^{n} \frac{ISC_t + SOM_t + EC_t}{(1+r)^t}}{\sum_{t=1}^{n} \frac{EO_t}{(1+r)^t}} \quad (1)$$

where ISC_t is the invested storage capital in year (t); SOM_t is the storage operation and maintenance costs in year (t); EC_t is the input energy cost (t); r is the annual discount rate (typically 10%); and EO_t is the value of released energy in year (t).

The LSC for the hydrogen storage technology will have additional revenue potential realized in the sale of both hydrogen and oxygen gases as a commodity. Equation (1) is therefore expanded to include $H2_t$ and $O2_t$, and the LSC is expressed as shown in Eq. (2):

$$LSC = \frac{\sum_{t=1}^{n} \frac{ISC_t + SOM_t + EC_t}{(1+r)^t}}{\sum_{t=1}^{n} \frac{EO_t + H2_t + O2_t}{(1+r)^t}} \quad (2)$$

where $H2_t$ is the value of sold hydrogen gas year (t) and $O2_t$ is the value of sold oxygen gas in year (t).

To evaluate the economic competitiveness of hydrogen energy storage systems utilizing the "surplus" or "grid constrained" renewable energy generation, several configurations are considered here to conclude the most economic scenario.

Scenario 1: Selling 100% of the hydrogen and oxygen gases produced by electrolyzer, and no electricity to sell (i.e., no fuel cell electricity generation).

Scenario 2: Selling 100% of the H_2 gas stored as electricity injected back to the power grid through the FC electricity generation (i.e., no H_2 gas to sell), and selling 100% of the O_2 gas.

Scenario 3: Selling 50% of the produced hydrogen as gas while the other 50% is sold as electricity to the grid via the fuel cell, and selling 50% of the oxygen as gas while the remaining 50% is vented to the atmosphere (i.e., not making use of half the produced oxygen value).

Scenario 4: Selling 100% of the hydrogen as gas (i.e., no fuel cell generation and no electricity to sell) and no selling of oxygen gas (i.e., not making use of all the produced oxygen value).

Scenario 5: Selling 100% of the stored hydrogen as electricity back to the power grid (i.e., no H_2 gas selling), and no O_2 gas selling (i.e., not making use of all the produced oxygen value).

Note that, Scenarios 4 and 5 do not utilize the by-product oxygen gas.

These five scenarios are tested on the “hydrogen office” energy storage system, as a case study, and the levelized cost per unit output is calculated for each scenario using Eq. (2). The Hydrogen Office, in Methil Docks Business Park in Scotland, employs a wind/hydrogen energy storage system that has been installed to demonstrate the potential of HES in storing surplus renewable energy. The Hydrogen Office's main components are shown in **Figure** 7, the Capital expenditure (CapEx) and Operational Expenditure (Opex) data are given in **Table 1**, and the market value for the by-product H_2 and O_2 gases is given in **Table 2** [34].

Figure 8 shows the LSC results for the five scenarios. It can be seen from figure that the most financially competitive configuration for the hydrogen energy storage technology is realized in Scenario 1. A favorable result is also seen in Scenarios 2 and 3. The least competitive configuration is seen in Scenario 5 when none of the gases is sold as commodity. Although hydrogen has a high financial value when sold as a gas, Scenario 4 demonstrates that it is not competitive when sold on its own.

It can be concluded from **Figure 8** that the hydrogen energy storage technology has a great potential and financial benefit in enabling the projected increase of renewable generation into the electrical network as it allows alternative economic pathways for

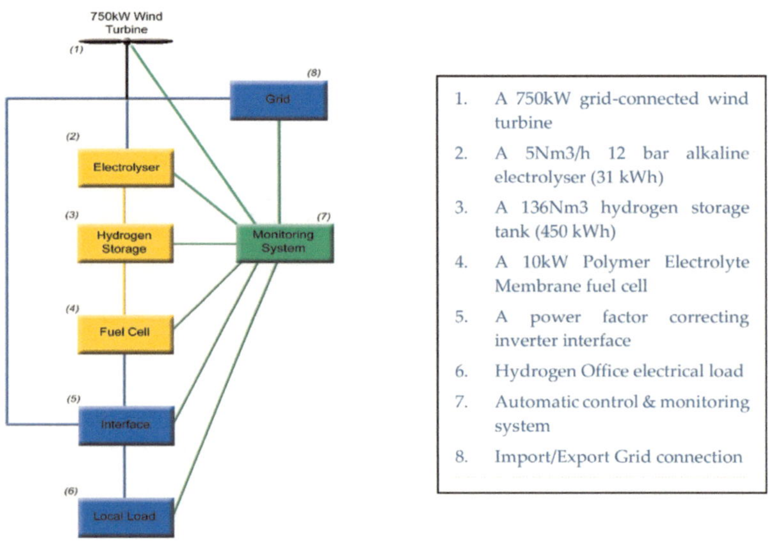

Figure 7.
The Hydrogen Office simplified system overview.

CapEx & OpEx data	
Hydrogen CapEx	
Electrolyzer (£/kW)	£ 2500.00
Storage (£/kWh)	£ 27.00
Fuel cell (£/kWh)	£ 4000.00
Hydrogen OpEX	
Electrolyzer (£/kW)	£ 50.00
Storage (£/kWh)	£ 2.00
Fuel cell (£/kW)	£ 100.00

Table 1.
HEST capital and operational expenditure data.

Market data	
O_2 Sale (£/Ton)	£ 3000.00
H_2 Sale (£/Ton)	£ 5000.00

Table 2.
Market value for hydrogen and oxygen gases.

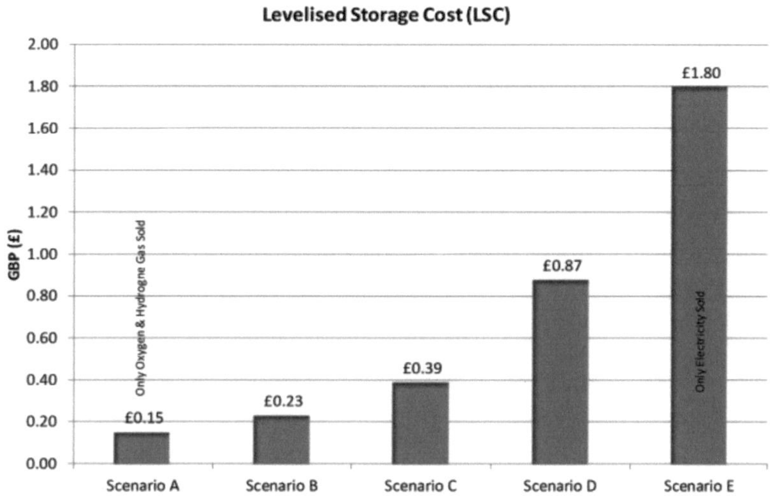

Figure 8.
The LCM simulated output costs for the HEST scenarios.

the surplus renewable generation. The stored hydrogen is not only limited for electricity production, but can also be sold for another end uses. Moreover, HEST has the most economic potential when its by-product oxygen is sold as well.

The HEST levelized cost is then compared to the levelized costs of other conventional energy storage technologies as obtained from a research conducted by NREL and summarized in **Table 3** [45]. **Figure 9** illustrates this comparison.

It can be seen from **Figure 9** that CAES and PHS are more cost competitive than the five HEST scenarios proposed. This shows that there is still a need for reducing the HEST CapEx or increasing its turn-around efficiency to increase its financial competitiveness. However, HEST can compete with NaS and RFB technologies when it is used in conjunction with the oxygen gas by-product selling. Additionally, HEST competes with NiCd battery technology when only 50% of the oxygen gas is sold and half the hydrogen is sold as gas and the other half as electricity. HEST is not competitive when used for only selling electricity (Scenario 5) or only selling H_2 gas (Scenario 4) without selling the O_2 gas.

5.2 Efficiency of renewable hydrogen energy storage systems (RHESS)

All energy storage systems have varying degrees of inefficiency (turn-around efficiency), with typical efficiency ranging from 45 to 80%. Hydrogen energy storage systems' efficiency can be considered higher especially when implemented with RES because of the following:

- The efficiency of electrolysis is high.
- They commonly utilize a fuel cell that has a conversion efficiency a lot higher than that of the combustion engine technology.

Technology	Levelized storage cost (LSC)
NiCd battery	₤ 0.52
NaS battery	₤ 0.16
Radox flow battery (RFB)	₤ 0.17
Pumped hydrostorage (PHS)	₤ 0.08
Compressed air energy storage (CAES)	₤ 0.06

Table 3.
Levelized costs of other energy storage technology [45].

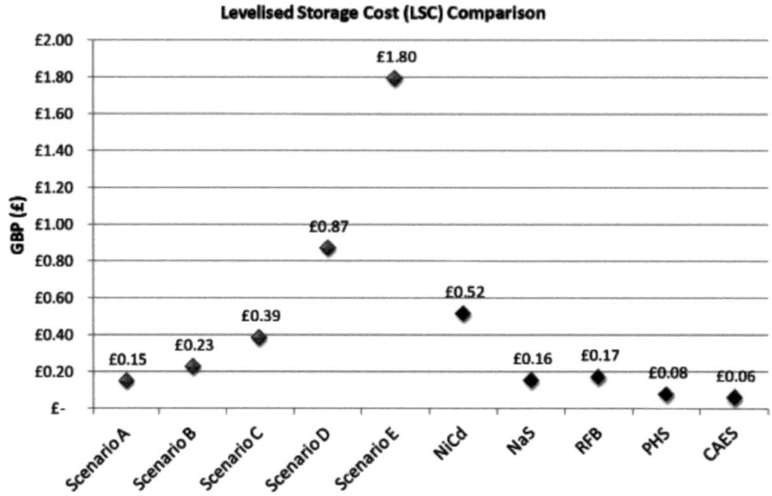

Figure 9.
The levelized cost of HEST versus the levelized costs of other conventional energy storage technologies.

- Their efficiency can be increased by utilizing the output heat from electrolyzers and fuel cells in process heating.

- When utilized with RES, they capture and store the excess in renewable energy that would have otherwise been dumped and this in turn adds to their efficiency.

- They do not only store the electrical energy for future re-use like all other conventional ESS, but also allow both hydrogen and oxygen gases to be sold as commodities thus increasing the system economic efficiency. The 3:1 increase in revenue options, shown in **Figure 6**, opens the potential for downstream applications like car fuelling, fertilizer production, and high and low-grade heat applications in addition to electricity.

6. Sizing a hybrid renewable hydrogen energy system (HRHES)

The available sizing models are either Commercial software (like EMCAS, EnergyPLAN, energyPRO, GTMax, IKARUS, Invert, MiniCAM, NEMS, ORCED, PERSEUS, ORCED, PERSEUS, PRIMES, ProdRisk, RAMSES, RETScreen, SIVAEL, STREAM, WASP, and WILMAR [46, 47]) or standalone modeling techniques. The commercially available software does not offer simulating the HEST as part of a hybrid renewable energy system, and those which can [like HOMER, BALMORAL, H_2RES, ENPEP-BALANCE, HYDROGEMS (incorporated into Transys16), SimREN, and UniSyD3.0] require a significant quantity of input data and substantial computing resources to perform well and still some of them are not capable of sizing the hydrogen energy system. The more advanced standalone modeling techniques, like genetic algorithms (GA), particle swarm optimization (PSO), and simulated annealing (SA) need significant computing resources.

To address the large input data requirements of the commercially existing software and the significant computing resources needed by advanced GA, PSO, and SA techniques, a new deterministic sizing methodology that offers a rapid initial system sizing of a hybrid renewable hydrogen energy system (HRHES) with the minimal amount of input data and computer resources is given here. This simple technical sizing technique, referred to as deterministic [48, 49], can provide a rapid and reasonably accurate system sizing [50] while using limited number of input data. This approach can therefore play an important role at the initial design phase of HRHES.

6.1 The developed deterministic sizing algorithm

To demonstrate the proposed new deterministic sizing methodology that offers a rapid initial system sizing of a hybrid renewable hydrogen energy system (HRHES) with the minimal amount of input data and computer resources and thus supports its initial design, a HRHES sizing model is developed here based on the presence of solar and wind as the renewable resources combined with HEST to meet the demands of an electrical load. The outline of the developed deterministic sizing algorithm is shown in **Figure 10** and is detailed in the following subsections.

6.1.1 Sizing the integrated renewable energy sources (wind turbine and solar photovoltaic)

The first criterion in sizing a HRHES is sizing the renewable energy sources (RES). The objective of including the RES generation into the sizing routine is to minimize the difference between the average load demand ($\overline{P}_{dem}$) and the average renewable energy generation ($\overline{P}_{gen}$). Typically, the duration of the analysis is 1 year to allow incorporating the seasonal variation of the demand and renewable generation. In the developed model, the wind turbine is considered as the primary renewable energy source and the photovoltaic (PV) as the secondary. Hence, the capacity factor for each technology is first determined.

The capacity factor (CF) is defined as the average power output ($\overline{P}$) from the renewable device as a percentage of the maximum power output (P). The capacity factor of a wind turbine (CF_{wind}) is given by Eq. (3), and that for a PV (CF_{PV}) is given by Eq. (4).

$$CF_w = \frac{\overline{P}_w}{P_w} \tag{3}$$

$$CF_{PV} = \frac{\overline{P}_{PV}}{P_{PV}} \tag{4}$$

where $\overline{P}_w$ and $\overline{P}_{PV}$ are the average power outputs from the wind turbine and the PV, respectively, while P_w and P_{PV} are their rated power outputs.

Given that sizing the wind turbine is usually restricted to the unit sizes available in the market, which in general are 3, 5, 6, 10, 15, 20, 50, 250, 330, 500, 850, 900,

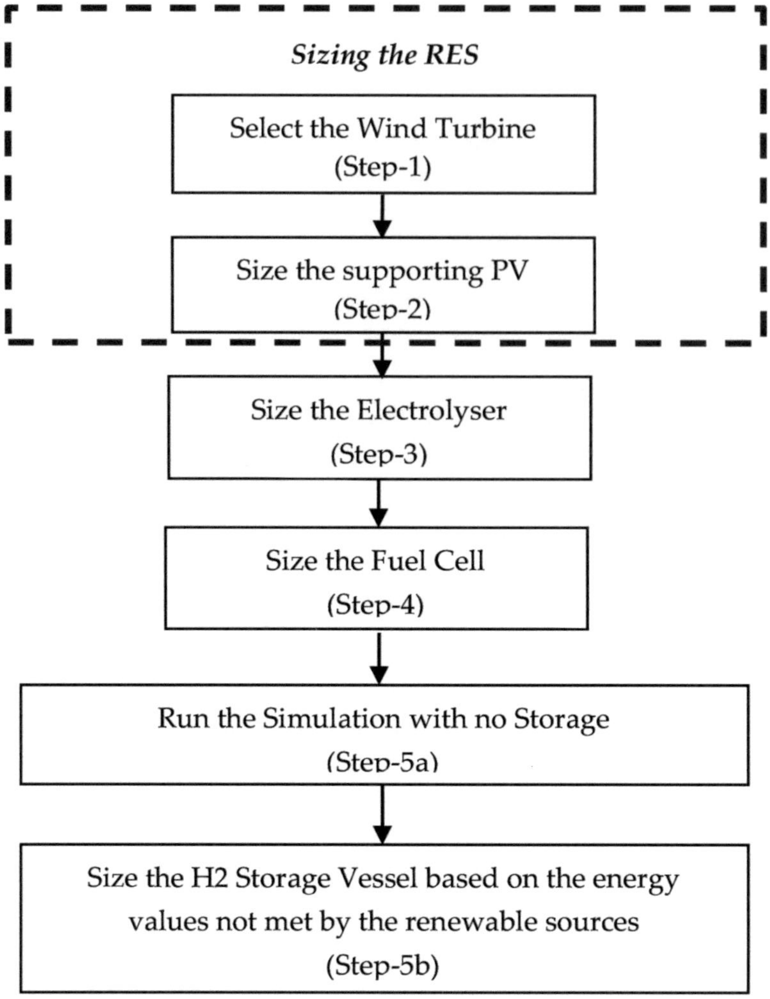

Figure 10.
The proposed deterministic sizing algorithm.

1200, 2200, 3200 kW, and thus the wind turbine is sized first. The average power of the wind turbine ($\overline{P}_w$) is first selected to be close to the average demand of the defined load ($\overline{P}_{dem}$); i.e., $\overline{P}_w \cong \overline{P}_{dem.}$

The rated power output for the wind turbine is then calculated using the proposed wind turbine model given by Eq. (5). Note that, the annual average site wind speed is calculated using Eq. (6) and the wind turbine rotor coefficient of performance (Cp), which is a measure of wind turbines blade rotor effectiveness at converting the power in the wind to mechanical power, is calculated using Eq. (7).

$$P_w = \left(\frac{1}{2} C_P \rho A v^3\right) \tag{5}$$

$$v = 1 + S_{var} \cos\left(t_{hr}\left(\frac{360}{8760}\right)\left(\frac{\pi}{180}\right)\right) exp\left\{\left(-\left(\frac{\overline{\mathrm{V}}}{\mathrm{C}}\right)^{\mathrm{k}}\right)\right\} \tag{6}$$

$$C_p(\lambda) = 0.5(\lambda - 0.02\beta^2 - 2.9)e^{\lambda} - 0.0303\lambda \tag{7}$$

where λ is the tip speed ratio, ratio of the wind speed, and the speed at which the wind turbines rotor tips are traveling, and it is found using Eq. (8); and B is the blade's pitch angle.

$$\lambda = \frac{\frac{\pi n D}{60}}{v} \tag{8}$$

where n is the turbine RPM; D is the turbine rotor diameter; and v is the wind speed.

The second step is to define the rated power of the supporting PV array (P_{PV}) using Eq. (9). In Eq. (9), P_{PV} is calculated by using the values obtained from Eqs. (3)–(5).

$$P_{PV} = \frac{\overline{P}_{dem} - (CF_w P_w)}{CF_{PV}} \tag{9}$$

6.1.2 Sizing the electrolyzer (hydrogen generator)

After sizing the RES, the size of the electrolyzer (P_{EL}) is calculated by subtracting the minimum load demand (P_{dem_min}) from the total rated power that can be delivered by the renewable energy sources. However, it has been found out that such a size can be very large and underutilized and because electrolyzers can be very expensive, it is desirable to operate them with a high level of utilization. Reducing this calculated value by around 50% [51], as shown in Eq. (10), increases the electrolyzer level of utilization but there will be times when the total renewable generation exceeds the total power that can be absorbed by the load and the electrolyzer of the RHESS.

$$P_{EL} = (P_{PV} + P_{wind} - P_{dem\_min})/2 \tag{10}$$

6.1.3 Sizing the fuel cell

When the load demand exceeds the renewable generation, this deficit is met by the fuel cell generation. The fuel cell converts the chemical energy of the stored hydrogen into electrical energy to supply the demand. The fuel cell size is selected to meet or exceed the load maximum power demand ($P_{dem_{MAX}}$). Typically, a margin of 20% is added to the size of the fuel cell to accommodate any modest increase in peak demands [50]. The fuel cell power output (P_{FC}) is therefore calculated as shown in Eq. (11).

$$P_{FC} \cong 1.2(P_{dem_{MAX}}) \tag{11}$$

6.1.4 Sizing the hydrogen storage tank

The RES are first sized to supply a specified load on an annual average basis using Eqs. (3)–(9). Sizes of the appropriate electrolyzer and fuel cell are then identified using Eqs. (10) and (11). Because there could be times when there is no or insufficient renewable generation to supply the demand, assessing the correlation of the load demand and renewable generation is therefore needed. Simulating the energy system without storage, as shown in **Figure 11**, using the load and calculated sizes for PV and wind turbine allows to identify the correlation between the load demand and renewable generation.

The difference between the load demand and the renewable generation at different timings can be found by subtracting the load demand from the renewable

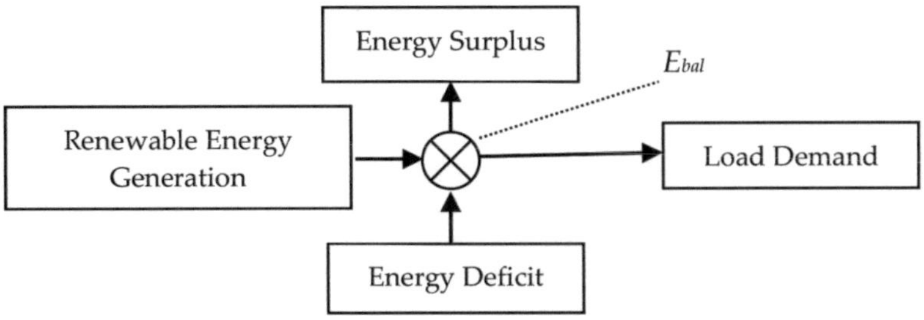

Figure 11.
Simulating the system without energy storage.

resource value for each recorded sample. Summing the differences for all the sample intervals yields a negative value, indicating a supply deficit, which defines the energy storage size. Equation (12) defines the size of energy storage (E_S) needed to cover this deficit.

$$\overline{E}_S = \sum E_{bal} < 0 \qquad (12)$$

The hydrogen storage tank must be sized to hold enough hydrogen for the fuel cell to deliver the energy requirements ($\overline{E}_S$), thus the average fuel cell conversion efficiency is considered. Therefore, the energy that is required to be stored within the hydrogen storage tank ($\overline{E}_{tank}$) can be defined as shown in Eq. (13).

$$\overline{E}_{tank} = \overline{E}_s . \frac{1}{\overline{\eta}_{FC}} \qquad (13)$$

where $\overline{\eta}_{FC}$ is the fuel cell conversion efficiency considering the lower heating value (LHV) of the hydrogen gas.

The average volume of hydrogen ($\overline{VL}_{tank}$) that needs to be stored in the hydrogen tank is then calculated, based on the absolute energy content of the hydrogen gas, using Eq. (14). The absolute energy content of the hydrogen gas is known as the higher heating value (HHV) of the hydrogen gas which is known to be 3.55 kWh/ Nm^3 [52].

$$\overline{VL}_{tank} = \frac{\overline{E}_{tank}}{3.55} \qquad (14)$$

7. Modeling the effect of thermal transients on the hydrogen production of renewably powered electrolyzers and utilizing the developed model as a tool to identify performance issues within operational hydrogen systems

A great challenge that faces the application of renewable-powered hydrogen energy storage systems is the ability to accurately determine the hydrogen production of an electrolyzer running on RES. Other challenges include their high costs and the need to guarantee their reliable and safe operation. A way forward toward achieving cost reduction is to lower their operation and maintenance costs by improving the hydrogen production efficiency and the system performance [53]. To achieve this, while ensuring safe system operation, the HES system must be able to handle the hydrogen gas securely and any leak must be quickly identified. If this is not the case, any leakage will impact the overall system efficiency and the operation cost and could result into a potential safety hazard as well.

To address these challenges, a model that includes the thermal compensation effect to accurately simulate the hydrogen production from an electrolyzer fed by RES is developed in this section. The developed thermally compensated electrolyzer model can also be used as a tool to detect any H_2 leakage and identify performance issues within an operational electrolyzer. The developed model, when implemented on working HES systems, allows the identification of hydrogen leakages without the need for maintenance inspection thus reducing the operating costs.

Ignoring the effect of temperature on the electrolyzer hydrogen production, as the case with HOMER, may lead to unrealistic simulation results because the electrolyzer production efficiency at lower temperatures is lower than that at full production temperature. This result has been illustrated when the developed model was implemented on a 30 kW electrolyzer, as a case study, to simulate the effect of temperature across its full range of hydrogen production. **Figure 12** demonstrates the impact of heat compensation on the electrolyzer hydrogen production efficiency.

The developed model is based on the combination of heat transfer theory, fundamental thermo-dynamics, and empirical electrochemical relationships, as measured from operating systems. The developed model is detailed in the following subsections.

7.1 The algorithm proposed for identifying the effect of thermal transients on the electrolyzer hydrogen production

To identify the effects of thermal transients on the overall hydrogen production from an electrolyzer, a three-step algorithm is developed as shown in **Figure 13**. The first step involves simulating the hydrogen production from a renewable-powered electrolyzer with the electrolyzer model compensated using the effects of temperature on its hydrogen production. The second step involves repeating the simulation, but with the electrolyzer temperature fixed at its full working temperature (i.e., effects of thermal transients ignored). In the final step, the overall effect of thermal transients on hydrogen production is calculated by subtracting the step 1 hydrogen

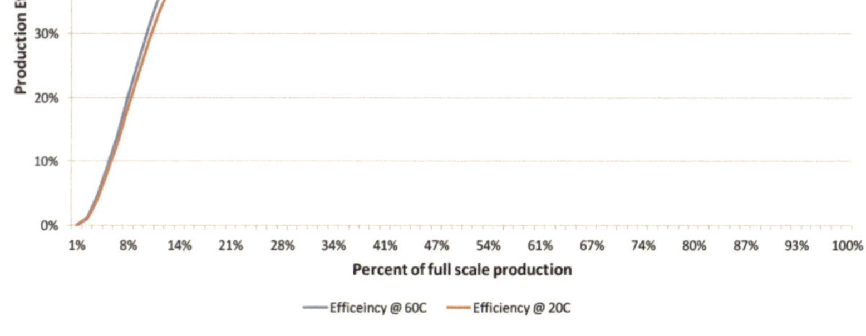

Figure 12.
Impact of heat compensation on the efficiency of a 30-kW electrolyzer.

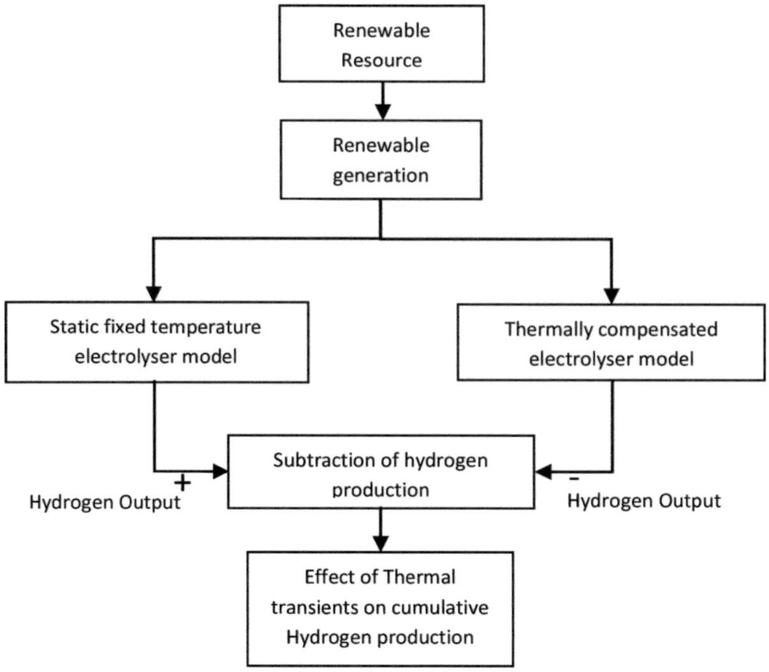

Figure 13.
Proposed algorithm to identify the impact of thermal transients on the electrolyzer hydrogen production.

output of thermally compensated model from the hydrogen output of the fixed temperature model of step 2.

7.2 Developing the electrolyzer model

A robust electrolyzer model, shown in **Figure 14**, is developed in this section to be used in the proposed algorithm to identify the effects of thermal transients on the overall hydrogen production. To formulate an accurate and robust electrolyzer model that can accurately predict the electrochemical and thermal dynamic behavior of an advanced alkaline electrolyzer, the model is developed based on Øystein Ulleberg model [19] while integrating the voltage/current U-I relationship, the faraday efficiency, as well as the thermal and the pressurized hydrogen storage modeling components.

7.2.1 The voltage/current U-I curve

An electrolyzer operating characteristic is determined by its voltage and current profile. The quantity of hydrogen produced by an electrolyzer varies with the amount of current passing through the electrolytic cell stack. The electrolytic cell voltage develops as more current is absorbed by the electrolyzer to increase the gas output flow. This U-I relationship would be a straight line for an ideal electrolyzer; however, it is a nonlinear relationship due to losses occurring in the electrochemistry and cell structure. The relationship is affected by the ohmic resistance of the electrolyte and electrodes as well as the parasitic loss of "stray" electrolysis. The parasitic loss of stray electrolysis is a phenomenon where the electrons flow down the electrolyte fluid channels instead of flowing directly between the electrodes themselves.

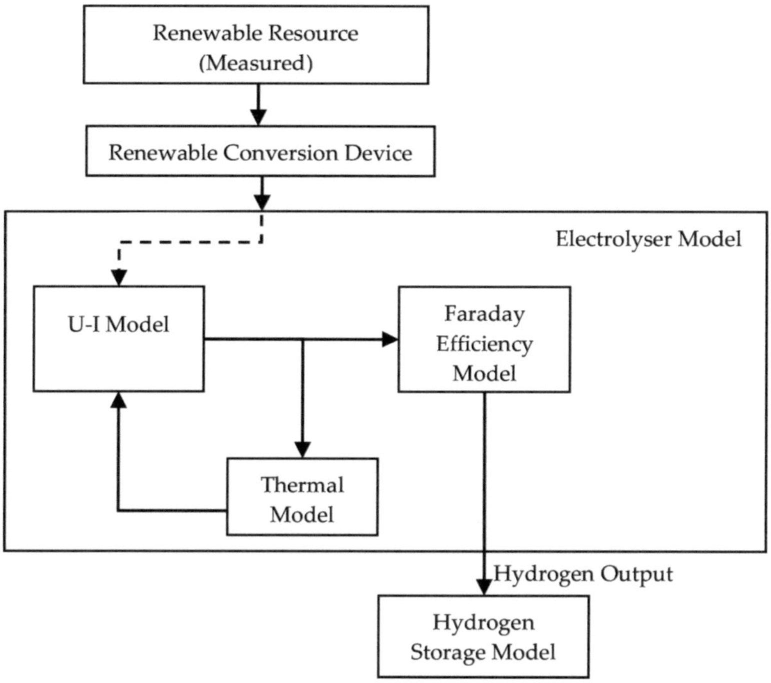

Figure 14.
Incorporating the effect of thermal transients in electrolyzer model.

The voltage (U) required to breakdown the water to produce hydrogen can be expressed in terms of (U_{rev}). The voltage required to facilitate the electrolytic dissociation of water molecules is temperature dependent and can be expressed as shown in Eq. (15).

$$U = U_{rev} + \frac{(r_1 + r_2 T)}{A} I + slog\left(\left(\frac{t_1 + \frac{t_2}{T} + \frac{t_3}{T^2}}{A} I\right) + 1\right) \quad (15)$$

where U is the water breakdown (or hydrogen production) voltage (V); U_{rev} is the overvoltage beyond reversible electrochemical cell voltage; $r_{1,2}$ is the empirical ohmic resistance parameter of electrolyte (Ωm^2); T is the temperature (K); $t_{1,2,3}$ is the empirical overvoltage parameter of electrode ($mA^{-1}\ m^2$); s is the overvoltage parameter of electrode (V); A is the electrode area (m^2); and I is the current (A).

The reversible cell voltage (U_{rev}) is calculated using the empirical Nernst equation for electrolysis given by Eq. (16) [20].

$$U_{rev,T(K)} = 1.5184 - 1.5421 \times 10^{-3} T + 9.523 \times 10^{-5} T lnT + 9.84 \times 10^{-8} T^2 \quad (16)$$

7.2.2 The Faraday efficiency

The Faraday efficiency is the ratio between the actual and maximum theoretical hydrogen mass that can be produced by an electrolyzer. Faraday efficiency losses are caused by parasitic current losses within the electrolysis cell stack. The parasitic current loss increases as a percentage of the overall current with the decreasing current densities and increasing temperatures. Therefore, the percentage of parasitic current loss to the total current flow increases with decreasing current densities. An empirical equation for the Faraday efficiency is shown in Eq. (17).

$$\eta_F = \frac{\left(\frac{I}{A}\right)^2}{f_1 + \left(\frac{I}{A}\right)^2} f_2 \tag{17}$$

where η_F is the Faraday efficiency; A is the electrode area (m^2); I is the current (A); f_1 is the Faraday efficiency parameter $mA^2\ cm^{-4}$; f_2 is the Faraday efficiency parameter (number between 0 and 1); and f_1 and f_2 are selected empirically.

Faraday's law also models the production rate of hydrogen in an electrolytic cell. The production rate of hydrogen is directly proportional to the transfer rate of electrons at the electrodes. This is equivalent to the electrical current provided by the power supply. Therefore, the total hydrogen production rate in an electrolysis stack consisting of several cells connected in series can be expressed, as shown in Eq. (18).

$$\dot{n}_{H_2} = \eta_F \frac{n_c I}{zF} \tag{18}$$

where $\dot{n}_{H_2}$ is the molar flow rate ($mol\ s^{-1}$); η_F is the Faraday efficiency; z is 2 (number of electrons transferred per reaction); I is the current (A); F is the Faraday constant 96,485 $C\ mol^{-1}$; and n_c is the number of series cells in electrolyzer cell stack.

7.2.3 The thermal model

The production of heat in an electrolyzer is primarily caused by electrical inefficiencies. The energy efficiency can be calculated from the thermo-neutral voltage (U_{tn}) and the cell voltage (U) using Eq. (19).

$$\eta_e = \frac{U_{tn}}{U} \tag{19}$$

where η_e is the energy efficiency; U_{tn} is the thermo-neutral voltage $\cong$ 1.477 V; and U is the cell voltage.

The value for U_{tn} remains almost constant within the pressure and temperature range considered here (0–1200 kPa pressure, 0–80°C temperature), this value is 1.477 V [21].

The operating temperature of an electrolyzer can be found from the overall thermal energy balance of the electrolysis system. The thermal energy balance of the electrolyzer can be expressed, as shown in Eq. (20), where Eq. (21) calculates the thermal energy created by the electrolysis process, and Eq. (22) is used to calculate the thermal losses of the electrolyzer system. Equation (23) is applied to maintain the electrolyzer temperature at or below the maximum temperature specified by manufacturer; it is assumed that electrolyzer cooling system is sufficient to remove the excess heat generated by the electrolysis process.

$$C_t \frac{dT}{dt} = \dot{Q}_{gen} - \dot{Q}_{loss} - \dot{Q}_{cool} \tag{20}$$

$$\dot{Q}_{gen} = n_c(U - U_{tn})I = n_c U I (1 - \eta_e) \tag{21}$$

$$\dot{Q}_{loss} = \frac{1}{R_t}(T - T_a) \tag{22}$$

$$\dot{Q}_{cool} > \dot{Q}_{gen} - \dot{Q}_{loss} \tag{23}$$

where $\dot{Q}_{gen}$ is the thermal energy created by electrolysis process; $\dot{Q}_{loss}$ is the thermal energy lost to the environment; $\dot{Q}_{cool}$ is the thermal energy dissipated by cooling system; C_t is the thermal capacity (or inertia) of electrolyzer (JK^{-1}); R_t is the thermal resistance of electrolyzer ($W^{-1}K$); n_c is the number of cells in the electrolysis stack; η_e is the energy efficiency (%); U_{tn} is the thermo-neutral voltage (V); U is the cell voltage (V); T is the electrolyzer temperature (K); T_a is the ambient temperature (K); and t is the time (seconds).

To calculate the electrolyzer temperature as time passes (T), it is assumed that the electrolyzer exhibits a constant heat generation and heat transfer profile for a small-time interval of not more than a few seconds. An intra-time-step steady-state thermal model can be expressed, as shown in Eq. (24), where T_{ini} is the initial temperature and Δ_t is the change in time.

$$T = T_{ini} + \frac{\Delta t}{C_t}\left(\dot{Q}_{gen} - \dot{Q}_{loss} - \dot{Q}_{cool}\right) \tag{24}$$

7.2.4 Pressurized hydrogen storage modeling

When hydrogen is produced by the electrolyzer, there is a need to store it and therefore there is a need to include hydrogen storage modeling to the proposed model. The two main components needed to model pressurized hydrogen storage is the formula for the pressure considering the gas behavior and the compressibility factor Z.

The ideal gas relationship can be used to describe the behavior of real hydrogen gas accurately only at relatively low pressures up to approximately 1450 psig and at normal ambient temperatures, results then become increasingly inaccurate at higher pressures. One of the easiest ways to account for this additional compression is through the addition of a compressibility factor, designated by the symbol Z. The Z factor is derived from the data obtained through experimentation and it depends on temperature, pressure, and on the nature of the gas. The Z factor is used as a multiplier to adjust the ideal gas law to fit into the actual gas behavior, as shown in Eq. (25).

$$P = Z\rho RT \tag{25}$$

where P is the absolute pressure in Pascal; ρ is the density; T is the absolute temperature in Kelvin; and R is the universal gas constant, 8.31434 Nm/mol K.

Calculating Z: The National Institute for Standards and Technology has developed a mathematical method for calculating compressibility factors accurately using a virial equation based on pressure (MPa) and temperature (K) [22]. The compressibility factor for hydrogen at different pressures and temperatures can be calculated to a high degree of accuracy by using Eq. (26) and the constants listed in **Table 4** [23, 24].

$$Z(P,T) = \frac{P}{\rho RT} = 1 + \sum_{i=1}^{9} a_i \left(\frac{100}{T}\right)^{b_i} \left(\frac{P}{1}\right)^{c_i} \tag{26}$$

where the equation and its constants are defined for pressures in Mega-Pascal (MPa) and temperatures in Kelvin (K).

7.3 Developing the electrolyzer model using MATLAB/Simulink

The previously developed modeling equations are used in this section to develop a MATLAB-Simulink model. The developed MATLAB model is then used in a novel way to assess the integrity of operational real-life hydrogen installations.

I	a_i	b_i	c_i
1	0.05888460	1.325	1.0
2	−0.06136111	1.87	1.0
3	−0.002650473	2.5	2.0
4	0.002731125	2.8	2.0
5	0.001802374	2.938	2.42
6	−0.001150707	3.14	2.63
7	0.9588528×10^{-4}	3.37	3.0
8	$-0.1109040 \times 10^{-6}$	3.75	4.0
9	0.1264403×10^{-9}	4.0	5.0

Molar mass: M = 2.01588 g/mol.
Universal Gas Constant: R = 8.314472 J/(mol K).

Table 4.
Constants to calculate Z.

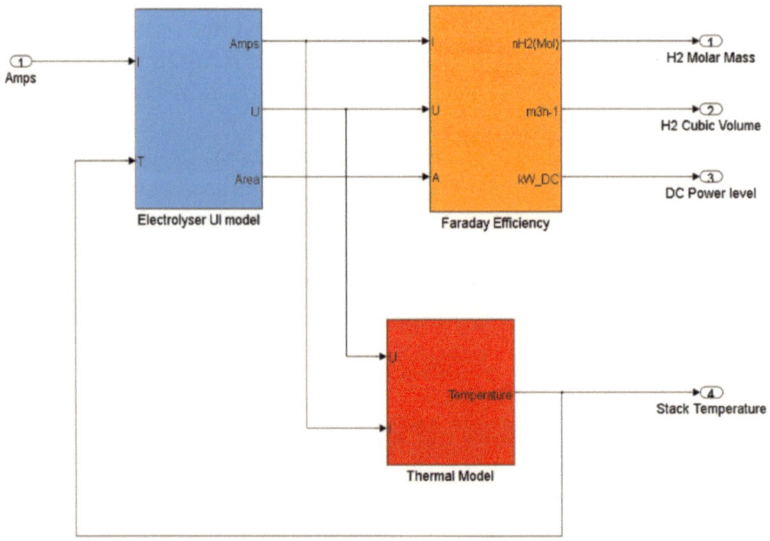

Figure 15.
Interaction between the U-I, the Faraday efficiency and the thermal subsystem models in MATLAB/Simulink.

The hydrogen generation and storage mathematical model, described by Eqs. (15)–(26), is implemented under the Simulink framework to develop the MATLAB Simulink model. **Figure 15** illustrates the interaction between the developed MATLAB Simulink subsystems for the U-I model, the Faraday efficiency model, and the thermal model but without any thermal compensation. The pressurized modeling is not shown in **Figure 15**, but it has been accounted for through the H_2 Molar Mass.

However, the simulation results for this model do not reflect the real hydrogen installations results, and thus the developed model is modified to include a newly added thermal compensation factor which is detailed in the next subsection.

7.4 Developing the thermally compensated electrolyzer model: compensating the temperature effect in an electrolysis system

Simulating an electrolyzer without considering the thermal compensation leads to a nonnegligible error in the simulation which can lead to a miscalculation of the return on investment (ROI). This is especially true during a cold start of an electrolyzer. An electrolyzer is said to be in a cold start when it is switched on in any of the following situations: (i) the electrolyzer is cold (not heated up and not at its standard operating temperature), (ii) not under pressure. Note that a standard operating temperature for an alkaline electrolyzer is about 80°C; for a proton exchange membrane (PEM) electrolyzer, it is about 70°C; and for a solid oxide electrolyzer (SOE), this varies with the material being used to construct the cells. In a cold start situation, the electrolyzer is cold and not under pressure and thus its efficiency is low as it requires pressurizing and heating itself up. This takes a short time if the electrolyzer is small; however, this time dramatically increases as the size of the electrolyzer increases. For alkaline electrolyzers, this time may be 1 s for small ones but can take up to several minutes for large ones; however, time is less for the PEM technology. The developed thermally compensated model will therefore be focused on alkaline electrolyzers as they are the ones that suffer the most from heat compensation effect.

To simulate the practical operation of an electrolysis system, it is important to include compensation of the effects of temperature on the electrolytic process. The exothermic thermal reaction that takes place during electrolysis in an alkaline electrolyzer impacts the energy efficiency of the gas generation process and especially the UI relationships. In other words, as hydrogen is being produced by the electrolyzer, an electrochemical reaction takes place at the electrodes. This reaction heats up the electrolyte and the associated electrode materials (this is known as the exothermic reaction) resulting into an increase in temperature which leads to reduction in the cell voltage and cell current needed to generate the hydrogen gas. In other words, the increase in the electrolyzer temperature reduces the power requirements of the electrolytic cells for the same hydrogen production, thereby increasing the system efficiency.

Integrating the thermal model within the previously developed electrolyzer model enables the thermal energy efficiency to be incorporated, and thus allows the model to generate output data that is extremely close to the real-world electrolyzer performance. **Figure 16** depicts the thermally compensated architecture included in the developed Simulink model. The new model considers the thermal energy generated by electrolysis (Q_{gen}), the thermal capacity of the electrolyzer itself to absorb and dissipate thermal energy (Q_{loss}), and the cooling system to maintain the thermal equilibrium required for an efficient hydrogen generation (Q_{cool}) and utilizes Eq. (23) to calculate the thermal balance of the electrolytic process.

7.5 Verification of the developed model

Two case studies are provided in this section to verify the developed model and validate that thermally compensated electrolyzer models are critical not only for designing new hydrogen installations, but also for monitoring the performance of operational ones. These case studies are carried out on two real-world field installed systems. The first one is used to verify that the developed model can accurately simulate hydrogen energy systems. The second is used to verify that the developed model can be used as a tool for investigating the operational integrity of operating electrolyzer systems and checking their performance while identifying any failures to support the reduction in maintenance requirements.

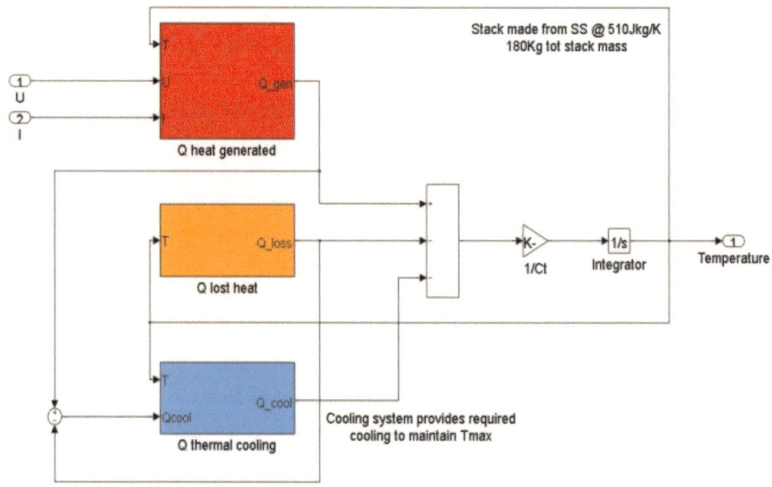

Figure 16.
The thermally compensated model developed under MATLAB/Simulink framework.

7.5.1 Verifying the developed model can accurately simulate hydrogen energy generators

A 30 kW real-world alkaline electrolyzer, which is operational within an existing hydrogen installation, is chosen to verify that the developed electrolyzer Simulink model can accurately simulate it. This 30 kW alkaline electrolyzer can develop 5.3 Nm^3/h of hydrogen at a pressure up to 1200 kPa. It consists of two electrolytic cell stacks; each stack has 90 cells configured in a series connected array. The electrolyzer is designed to operate at a temperature of 60°C. **Figure 17** shows the electrolyzer (i.e., the hydrogen generator) connected to a 4800 L gas bottle array for the storage of the generated hydrogen at a pressure up to 1200 kPa.

Table 5 gives the values for the electrolyzer variables, while values for the variables of the modular hydrogen storage system which is connected directly to the electrolysis system are given in **Table 6**.

A data-log of the current consumed by the real-world electrolyzer while in operation is given in **Figure 18**. The electrolyzer temperature and pressure responses to the current are also recorded and compared to the results from the developed Simulink model, as illustrated in **Figures 19** and **20**, respectively. Both figures demonstrate that the developed model results are very close to the data collected from the operating electrolyzer, thus confirming that the developed model can be used for accurately simulating real-world installations.

On further analyzing **Figure 19**, it can be noticed that the electrolyzer is switched on at time 133 s (cold start) and it reaches its operating temperature (and pressure) at time 309 s; thus, the time taken from the cold start to the operating temperature is 176 s. This means that the electrolyzer almost took 3 min to reach its operating conditions and substantial amounts of hydrogen will not be produced

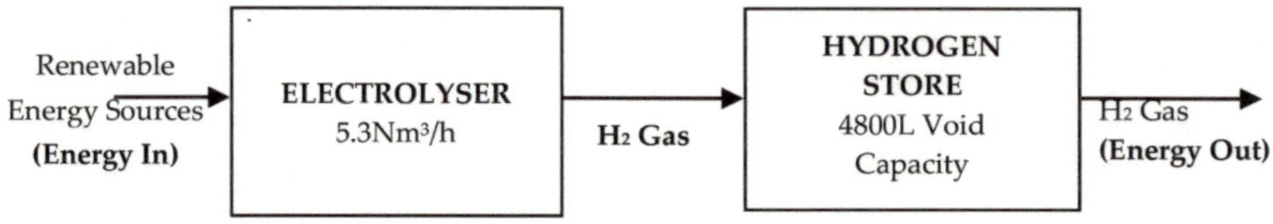

Figure 17.
Hybrid renewable H_2 generation and storage.

Variable	Description	Unit	Value
r_1	Electrolyte ohmic resistive parameter	Ωm^2	0.0000805
r_2	Electrolyte ohmic resistive parameter	Ωm^2	−0.0000025
A	Electrode area	m^2	0.37
S	Overvoltage parameter of electrode	V	0.19
t_1	Empirical overvoltage parameter of electrode	$A^{-1} m^2$	1.002
t_2	Empirical overvoltage parameter of electrode	$A^{-1} m^2$	8.424
t_3	Empirical overvoltage parameter of electrode	$A^{-1} m^2$	247.3
f_1	Faraday efficiency parameter	$mA^2 cm^{-4}$	200
f_2	Faraday efficiency parameter	0....1	0.94
V_{std}	Volume of ideal gas at STP	$m^3 mol^{-1}$	0.0224136
R_t	Thermal resistance of electrolyzer	$W^{-1}K$	0.018
C_t	Thermal capacity of electrolyzer	JK^{-1}	300,000
n_c	Number of cells in electrolysis stack	N	180
T_a	Ambient temperature	°C	20
T_{max}	Maximum operating temperature of electrolyzer	°C	60
T_{hyst}	Cooling hysteresis thermal band	°C	3

Table 5.
Electrolyzer variables—variables for the 30 kW alkaline electrolyzer.

Variable	Description	Unit	Value
V	Tank volume	m^3	4.8
Ta	Ambient temperature	°C	20

Table 6.
Variables for the hydrogen storage—4800 L void capacity.

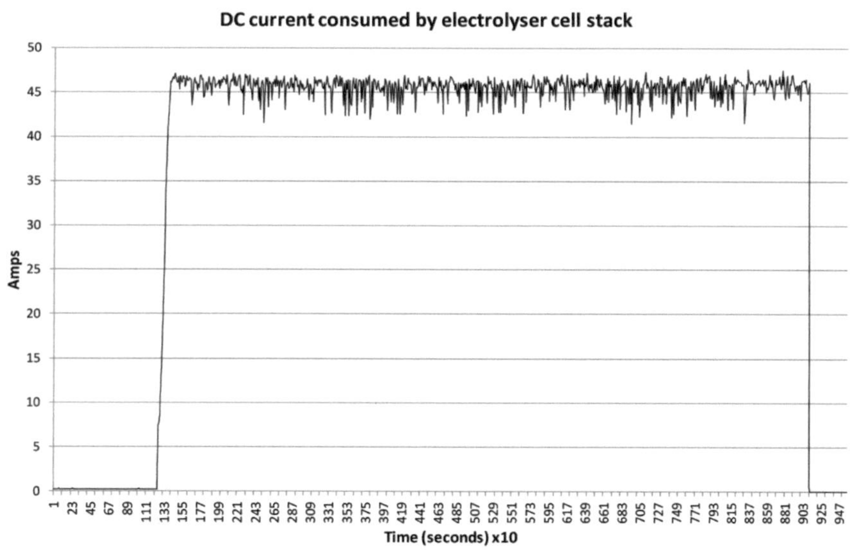

Figure 18.
Data-log of the current consumed by 30-kW electrolyzer while in operation.

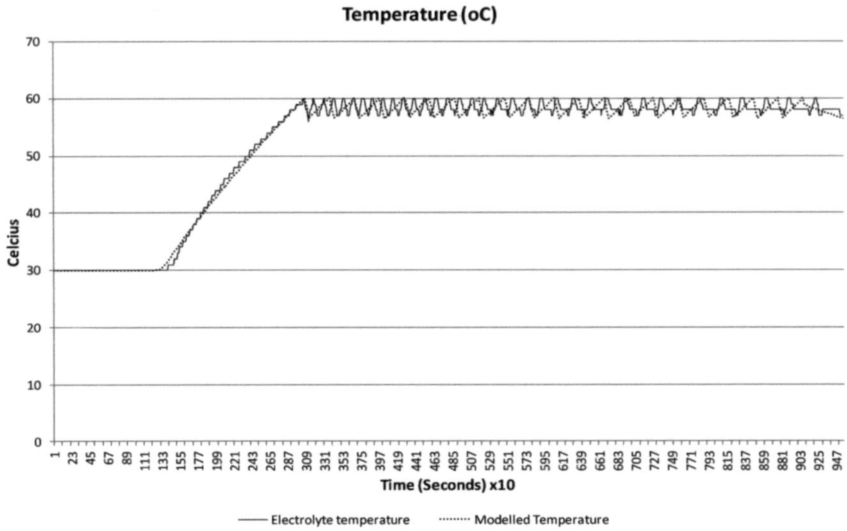

Figure 19.
Recorded electrolyzer temperature versus model output.

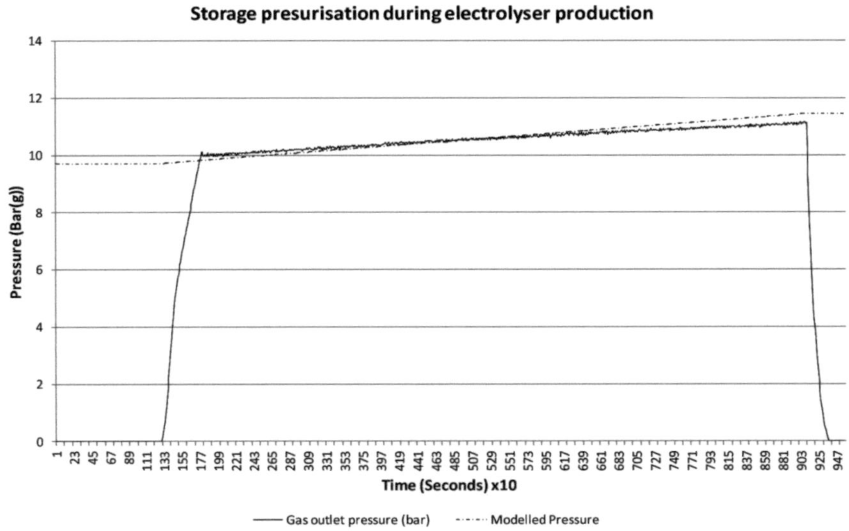

Figure 20.
Recorded electrolyzer pressure versus model output.

during this time. If such a small electrolyzer took 3 min to reach its steady-state operation, then it can be tangibly assumed that this time will be much higher for a larger scale electrolyzer and substantial loss in hydrogen production could be realized. Considering the financials, it will be consequently affected by the loss in the hydrogen production during the cold start period. It can therefore be concluded that the inaccurate hydrogen production numbers generated from nonthermally compensated hydrogen generators simulation models will generate misleading higher ROI values. Thus, it can be also concluded that the developed thermally compensated simulation model is essential for accurately calculating the potential for financial return of a hydrogen system since it allows the accurate computation of hydrogen production.

7.5.2 Verifying that the developed model can be used as a tool for identifying performance issues within operational hydrogen systems

The thermally compensated electrolyzer model is further tested in one of the most unexploited applications for any model—its use in a postinstallation scenario. When an electrolyzer model is developed, it is usually used in the preinstallation stage to investigate if the planned system will operate as anticipated when installed in the field. In this section, the developed model is further used in a postinstallation scenario to demonstrate that it is also capable to detect issues within operating systems removing the need for maintenance crew on-site inspection.

The developed Simulink model was used to simulate an operating hydrogen generator when its performance was detected to be not as anticipated for a couple of weeks. It was suspected that the electrolyzer has developed an internal issue, so the aim was to examine if the developed model can be used in a post-installation situation to determine this performance issue. The operating hydrogen system, on which the developed Simulink model was tested in a postinstallation scenario, is been operating in Africa for over 8 years and it employs a 30 kW alkaline electrolyzer identical to that given in **Table 5** connected to a storage system of 2499 L void volume capacity. On comparing the model output results to the on-site collected data, two performance issues were doubted. The first was an early degradation of the stack; however, this was disregarded as none of the other similar installed stacks illustrated such a drastic performance issue. The second was the presence of a hydrogen leak; and this was clear from the divergence between the modeled and recorded data shown in **Figure 21**. The figure shows that there was more H_2 production in the model results than what was achieved in the practical installation, and this in turn suggests a leak within the installed system. This suggestion was confirmed by an on-site inspection of the hydrogen system which revealed that a fitting in the pipe that carries the H_2 gas from the electrolyzer to the storage system had developed a premature failure. The fast rate of detected leak spray bubbling, shown in **Figure 21**, indicates the presence of a leak on two sides of the faulty pipe fitting shown in **Figure 22**. This finding clearly demonstrates the apparent benefit of the developed model in identifying leakages during operation,

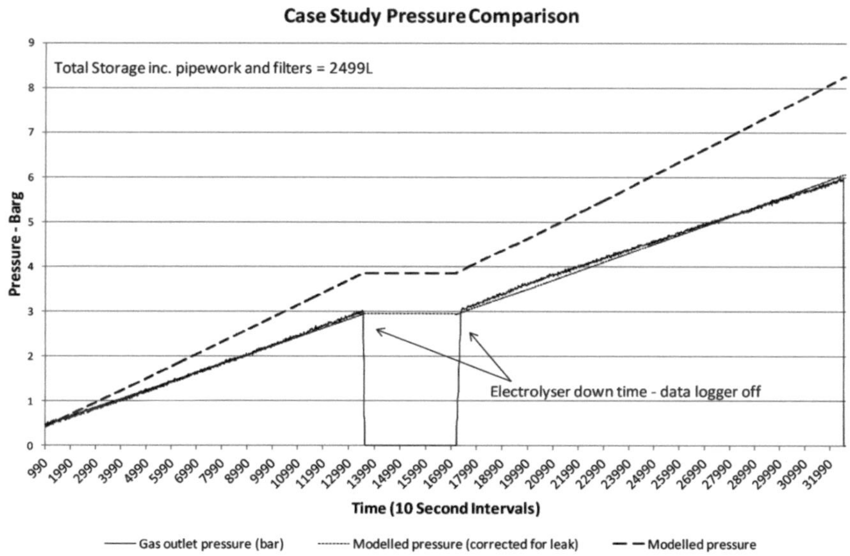

Figure 21.
Divergence between the modeled and recorded pressures indicating a suspected hydrogen gas leak.

Figure 22.
Source of leakage identified after on-site inspection.

and thus it can save the time and cost of maintenance inspection as well as preventing the safety hazards associated with the hydrogen vented in atmosphere.

The developed thermally compensated model has been able to reveal this hydrogen gas leak, which was about 10.89 g, equating to a 2.3% reduction in the overall system efficiency. This leak was so small for the leak detection system to detect; therefore, if many small leaks like this occur at different locations of a large-scale system without being detected by the safety alarm system this could lead to more financial losses and potentially a hazardous situation. Therefore, the developed model can be used as a tool to provide an early warning of leakages or other issues, and thus provided an extra layer of safety and a potential for increasing the financial return through the development of a predictive maintenance system.

8. Conclusion

In conclusion, the contributions to knowledge that has been presented within this chapter can be summarized as follows:

- A novel levelized cost model has been developed for investigating the financial competitiveness of the hydrogen energy storage technology. It has been identified that hydrogen use as an energy storage mechanism achieves the most financial competitiveness when the by-product oxygen is utilized.

- A new deterministic sizing methodology that offers a rapid initial sizing of renewable hydrogen energy storage systems has been given. The proposed method requires a very limited number of input data to offer an initial system size for a hybrid renewable hydrogen energy storage system (HRHES) very quickly, and thus it is useful at the very early initial design phase to assist in the early decision-making for system implementation. To develop this sizing model, a model has been developed for every single item in the proposed HRHES (the implemented renewable energy sources, the electrolyzer, H_2 storage, and fuel cell). These models were then integrated together.

- An algorithm for modeling the impact of thermal transients, especially in alkaline electrolyzers, on the overall hydrogen production has been developed. The prolonged thermal transients, associated with electrolyzers fed by renewable energy sources, result into extended periods of time where the electrolyzer does not produce hydrogen at its highest efficiency, and thus resulting into an overall reduction in its hydrogen production. The typical effect of thermal transients on the electrolyzer hydrogen production can be found by using the proposed algorithm, and a reduction in the cumulative hydrogen production was found to be in the range between 1 and 3%.

- The thermally compensated electrolyzer model has been developed in Simulink and has proven, through a case study, to be able to accurately simulate hydrogen generation and storage systems. The developed model presents a key finding for the hydrogen industry as it does not only allow the investigation of hydrogen systems performance in a preinstallation scenario prior to embarking into the expensive capital investment, but also proven to be useful in postinstallation scenarios. The developed model was found to be able to simulate operational installed hydrogen systems and assist in identifying their performance issues accurately.

Author details

Dallia Mahmoud Morsi Ali
Robert Gordon University, Aberdeen, Scotland, United Kingdom

*Address all correspondence to: dolly.ali@hotmail.co.uk; d.ali@rgu.ac.uk

References

[1] Hydrogen Council Vision. How Hydrogen Empowers the Energy Transition. 2017. Available from: http://hydrogencouncil.com/wp-content/uploads/2017/06/Hydrogen-Council-Vision-Document.pdf

[2] Stern LN. The Economics of Climate Change: The Stern Review. Cambridge: Cambridge University Press; 2006

[3] The Importance of Flexible Electricity Supply. United States Department of Energy, Solar Technologies Programme, Solar Integration Series 1 of 3, DOE/GO-102011-3201; 2011

[4] Wilson IAG, McGregor PG, Infield DG, Peter J. Grid-connected renewables, storage and the UK electricity market. Hall, Elsevier Renewable Energy. 2011;**36**: 2166-2170

[5] Nelson DB, Nehrir MH, Wang C. Unit sizing and cost analysis of stand-alone hybrid wind/PV/fuel cell power generation systems. Elsevier Renewable Energy Journal. 2005;**31**

[6] Loisel R. Power system flexibility with electricity storage technologies: A technical–economic assessment of a large-scale storage facility. Electrical Power and Energy Systems. 2012;**42**: 542-552. DOI: 10.1016/j.ijepes.2012.04.058

[7] Orecchini F, Santiangeli A. Beyond smart grids - the need of intelligent energy networks for a higher global efficiency through energy vectors integration. International Journal of Hydrogen Energy. 2011

[8] Wade NS, Taylor PC, Lang PD, Jones PR. Evaluating the benefits of an electrical energy storage system in a future smart grid. Journal of Energy Policy. 2010

[9] McDowall J. Integrating energy storage with wind power in weak electricity grids. Journal of Power Sources. 2006;**162**:959-964

[10] Edberg O, Naish C, McNaught C. Energy Storage and Management Study. AEA Technology for the Scottish Government, ELL/000/077, revision 4; 2010

[11] Luickx PJ, Delarue ED, D'haeseleer WD. Considerations on the backup of wind power: Operational backup. Journal of Applied Energy. 2008

[12] Coll-Mayor D, Paget M, Lightner E. Future intelligent power grids: Analysis of the vision in the European Union and the United States. Elsevier Energy Policy Journal. 2006

[13] Constable J, Moroney L. High Rewards for Wind Farms Discarding Electricity 5-6th April 2011, Renewable Energy Foundation Information Note; 2011

[14] U.S. Energy Information Administration Office of Integrated Analysis and Forecasting. International Energy Outlook 2010. DOE/EIA-0484 (2010); 2010

[15] Daima TU, Li X, Kim J, Simms S. Evaluation of energy storage technologies for integration with renewable electricity: Quantifying expert opinions. Elsevier Environmental Innovation and Societal Transitions. 2012;**3**:29-49

[16] Beaudin M, Zareipour H, Schellenberglabe A, Rosehart W. Energy storage for mitigating the variability of renewable electricity sources: An updated review. Elsevier Energy for Sustainable Development. 2010;**14**: 302-314

[17] Gazey R, Ali D, Aklil D. Techno-economic assessment of hydrogen

energy storage systems for Enabling the projected increase of renewables onto electrical power grids. In: Proceeding of the IET Renewable Power Generation Conference RPG. Edinburgh, UK; 2011. Available from: http://conferences.theiet.org/rpg/index.cfm

[18] IEA Agreement on the Production and Utilization of Hydrogen. IEA-HIA Task 18 Sub-Task B Final Report; 2007. ISBN 978-0-9815041-0-0

[19] Gray D. Smart Grids and Virtual Power Plants: Point of Departure. Robert Gordon University; 2011

[20] Tessenderlo Group. Annual Report; 2003

[21] Maisonnier G, Perrin J, Steinberger-Wilckens R. Industrial surplus hydrogen and markets and production, deliverable 2.1 and 2.1a, Roads2HyCom, Document Number: R2H2006PU.1; 2007

[22] Crockett RGM, Newborough M, Highgate DJ. Electrolyser based energy management: A means for optimizing the exploitation of variable renewable energy resources in standalone applications. Solar Energy Journal. 1997; **61**(5):293-302

[23] Aklil TD, Goodhand R, Eguizábal MG. The h2seed Project. M. San Antonio Fuel Cell Seminar & Exposition; 2007

[24] Stoyel J. Hydrogen Mini Grid System: An Iconic Development for Yorkshire Forward

[25] Kraemer M. Renewable energy storage and hydrogen energy. In: Presentation to Grove Fuel Cell Symposium, London; 2009

[26] Ulleberg Ø, Nakken T, Eté A. The wind/hydrogen demonstration system at Utsira in Norway: Evaluation of system performance using operational data and updated hydrogen energy system modelling tools. International Journal of Hydrogen Energy. 2010; **35**(2010):1841-1852

[27] Hydrogen office opens in Scotland. Fuel Cells Bulletin. 2011;**2011**(2):9-10. DOI: 10.1016/S1464-2859(11)70092-X

[28] Yilanci A, Dincer I, Ozturk HK. A review on solar-hydrogen/fuel cell hybrid energy systems for stationary applications. Elsevier Progress in Energy and Combustion Science. 2009;**35**:231-244

[29] Gahleitner G. Hydrogen from renewable electricity: An international review of power-to-gas pilot plants for stationary applications. International Journal of Hydrogen Energy. 2013;**38**: 2039-2061

[30] Stiller C, Stubinitzky A. Hydrogen for grid scale storage of renewable energy. In: World Hydrogen Energy Conference. 2012

[31] Water electrolysis and renewable energy systems. Fuel Cell Today. 2013

[32] Forsberga P, Karlström M. On optimal investment strategies for a hydrogen refueling station. International Journal of Hydrogen Energy. 2007;**32**:647-660

[33] Aklil-D'Halluin D, Johnson E, Gazey R, et al. Remote Communities & Islands–An Economical Case for Hydrogen Technologies. UNIDO; 2008

[34] Ali D. Energy capacity and economic viability assessment of the renewable hydrogen energy storage as a balancing mechanism in addressing the electric system integration issues inherent with variable renewable energy resources. In: Proceeding of the IET Reliability of Transmission Distribution Networks Conference RTDN. London, UK: Dexter House; 2011. Available from: www.theiet.org/rtdn.http://www.bjadks.com/iet.tv/detail/2751.html

[35] Díaz-González F, Sumper A, Gomis-Bellmunt O, Villafáfila-Robles R. A review of energy storage technologies for wind power applications. Elsevier Renewable and Sustainable Energy Reviews. 2012;**16**:2154-2171

[36] Zheng J, Liu X, Xu P, Liu P, Zhao Y, Yang J. Development of high pressure gaseous hydrogen storage technologies. International Journal of Hydrogen Energy. 2012;**37**:1048-1057

[37] Chen H, Cong TN, Yang W, Tan C, Li Y, Ding Y. Progress in electrical energy storage system: A critical review. Progress in Natural Science. 2009;**19**: 291-312

[38] Ferreira HL, Garde R, Fulli G, Kling W, Lopes JP. Characterisation of electrical energy storage technologies. Energy. 2013;**53**:288-298

[39] Ibrahima H, Ilincaa A, Perron J. Energy storage systems characteristics and comparisons. Renewable and Sustainable Energy Reviews. 2008;**12**: 1221-1250

[40] Electric Energy Storage Technology Options: A White Paper Primer on Applications, Costs and Benefits. Palo Alto, CA: Electric Power Research Institute (EPRI); 2010. p. 1020676

[41] MacDonald M. UK Electricity Generation Costs Update. Department of Energy & Climate Change; 2010

[42] Akhil AA, Huff G, Currier AB, Kaun BC, Rastler DM, Chen SB, et al. DOE/EPRI 2013 Electricity Storage Handbook in Collaboration with NRECA. Sandia National Laboratories, SAND2013-5131; 2013

[43] Harrison K. Analysis of hydrogen and competing Technologies for Utility-Scale Energy Storage. In: IPHE Workshop. Spain; 2012

[44] Sundararagavan S, Baker E. Evaluating energy storage technologies for wind power integration. Solar Energy. 2012;**86**:2707-2717

[45] Steward D, Saur G, Penev M, Ramsden T. Ramsden. Cost Analysis Highlights Hydrogen's Potential for Electrical Energy Storage. Technical Report NREL/TP-560-46719; 2009

[46] Erdinc O, Uzunoglu M. Optimum design of hybrid renewable energy systems: Overview of different approaches. Renewable and Sustainable Energy Reviews. 2012;**16**:1412-1425

[47] Connolly D, Lund H, Mathiesen BV, Leahy M. A review of computer tools for analysing the integration of renewable energy into various energy systems. Applied Energy. 2010;**87**:1059-1082

[48] Castaneda M, Fernandez LM, Sanchez H, Cano A, Jurado F. Sizing methods for stand-alone hybrid systems based on renewable energies and hydrogen. In: Electro-Technical Conference (MELECON), 2012 16th IEEE Mediterranean. DOI: 10.1109/MELCON.2012.6196558

[49] Watson SJ, Ter-Gazarian AG. The optimization of renewable energy sources in an electrical power system by use of simulation and deterministic planning models. International Transactions in Operational Research. 1996;**3**(3/4)

[50] Wang C, Nehrir MH. Power management of a stand-alone wind/photovoltaic/fuel cell energy system. IEEE Transactions on Energy Conversion. 2008;**23**(3)

[51] Ulleberg Ø. Stand-alone power systems for the future: Optimal design, operation & control of solar-hydrogen energy systems. [Ph.D. dissertation]. Trondheim, Norway: Norwegian Univ. Sci. Technol.; 1998

[52] Gazey RN. Design and development of an integrated prototype renewable hydrogen energy system. [MPhil Thesis]. Robert Gordon University; 2006

[53] Ali D, Gazey R. Developing a thermally compensated electrolyser model coupled with pressurized hydrogen storage for modelling the energy efficiency of hydrogen energy storage systems and identifying their operation performance issues. Renewable & Sustainable Energy Reviews Journal. 2016;**66**:27-37